K.I.S.S. IT
TEAS® 6 MATH
EDITION

Jakara Lewis

Master of Science Education

Lewis & Wilder Publishing LLC

ATI TEAS® is a registered trademark of Assessment Technology Institute which does not sponsor or endorse this product.

Special thanks to my friend, Shank, for helping me format the entire book from cover to cover. This definitely would not have been possible without you!

Daytona Beach, Florida

www.lewis-wilderpublishing.com

Dedication

To

my biggest fan

The love of my life

Thank you for your never-ending support

unconditional love and strength

I love you

Mom

To

my intellectual twin

The reason for my hustle

Thank you for blessing me with your mind

keeping me grounded and reminding me to keep a smile

I love you

Dad

To

my better half

The reason for my smile

Thank you for helping me become a better person

pushing me to be great and loving me without prejudice

I love you

Allen

Without you all

This would not have been possible

Jakara

Acknowledgements

First and foremost, I would like to thank God for blessing me with this amazing gift that allows me to help others turn their dreams into reality. I have truly been blessed with some amazing and supportive individuals throughout my life, and for that I am beyond thankful. If not for your guidance, I wouldn't be who I am today.

To my mother, Adrianne, thank you for giving me life; I owe my entire existence to you. Thank you for supporting and inspiring me to be the best version of myself my entire life. You've always encouraged me and reminded me to keep pushing even when I wanted to give up. I could not have asked for a better mother; the love you have given me is without a doubt the reason why I am successful. The sacrifice and strength you've shown has always and will always push me to be better than I was yesterday. For this, I am forever thankful; I love you more words can describe.

To my grandmother, Gloria, thank you for all that you've contributed to my life; I am forever indebted to you. Thank you for providing love and comfort during the times I needed it the most. I've always been a dreamer, and I am so happy that you've been able to watch me grow into such an amazing woman. For this, I am forever thankful; I love you more than you'll ever know.

To my father, Kendrick, thank you for giving me life; I owe my entire existence to you. Thank you for supporting and inspiring me to be the best version of myself. Although things haven't always been easy, the tough times helped me to realize the importance of growth and forgiveness. We are truly one in the same, and I am thankful that God blessed me with you. This book would not have been possible without my intellectual twin, and for that I love you forever.

To my godmother, Sabrina, thank you for the unconditional love you've shown me throughout the years. Thank you for keeping me grounded and reminding me to always keep pushing no matter how hard life gets. I will carry your love and wisdom with me forever, I love you.

To all of the young ladies who hold a special place in my heart, thank you all for your love and friendship. Without you all continuously motivating me and reminding me of how amazing I am, this book would not have been possible. I thank God for blessing me with the few smart, beautiful, and ambitious young ladies I have the pleasure of calling my sisters and my best friends. I love you all more than words can describe.

To my better half, Allen, thank you for showing me the true definition of love. Life with you has been nothing short of amazing. Your protective nature, compassion for others, and fierce leadership skills are a few of the reasons why I love you. You have helped me to grow into a bigger, better, and fiercer version of myself. You see things in me, that I cannot see in myself and for this I am forever thankful. You stood by me every single day of this process, and I don't know if I can ever repay you for that. Just know it will always be me and you against the world, I GOT YOU! Thank you for loving me without prejudice, for without you this book would not have been possible. I love you, forever.

Test Breakdown

36 questions given　　32 questions scored

54 minutes allotted

4-function calculator provided

Topics

Numbers and Algebra (23 questions)

Apply Estimation Strategies and Rounding Rules to Real-World Problems

Convert Among Non-Negative Fractions, Decimals, and Percentages

Compare and Order Rational Numbers

Solve Real-World Problems Involving Ratios and Rates of Change

Solve Real-World Problems Involving Proportions

Solve Real-World Problems Involving Percentages

Perform Arithmetic Operations with Rational Numbers

Solve Real-World One or Multi-Step Problems with Rational Numbers

Translate Phrases and Sentences into Expressions, Equations, and Inequalities

Solve Equations in One Variable

Measurement and Data (9 questions)

Convert Within and Between Standard and Metric Systems

Explain the Relationship Between Two Variables

Evaluate the Information in Tables, Charts, and Graphs Using Statistics

Interpret Relevant Information from Tables, Charts, and Graphs

Calculate Geometric Quantities

Test Taking Strategies

1. DON'T OVERTHINK THE TEST

It's just a test! Like many, you've probably felt totally defeated by math your entire life. Fortunately, this study guide will help to ease your anxiety, but you must be willing to put in the work. The better you get, the more confident you will become!

2. MASTER THE CONTENT

Take your time and learn the material. Go through every example and complete all of the practice questions to help you retain the content. Practice these questions over and over until you master the concepts. For many of you, this will require hours and hours of studying for an extended period and that's ok! Studying should also be goal oriented; studying for hours means nothing if you haven't showed any signs of improvement. Practice until you can answer these questions confidently and correctly.

3. MANAGE YOUR TIME

Time management is one of the most important aspects of studying. Too often, people make the mistake of trying to "cram" the information into their minds in a short period of time. The reality is that this short cut rarely proves to be successful for test takers. Utilize the schedules that have been provided on the next page, and plan accordingly. If math is one of your major weaknesses, then you should plan to dedicate a few hours each day to study.

4. BE HONEST WITH YOURSELF

As you begin to study, make sure you always remain honest with yourself. Remember, one of the most important aspects of studying is progressing. The only way to progress is to work hard, study, and learn from your mistakes. Don't trick yourself into thinking you're comfortable with the content when you're truly not.

5. KEEP TESTING

As you begin studying, keep in mind the importance of continuously assessing yourself. Take as many practice tests as you can before you take the official test. Practice tests are great tools that allow you to mimic the testing conditions and expose you to a variety of ways in which the topics can be presented. Make use of the answer explanations and learn from your mistakes. Make sure you go back and review the content covering the questions that you missed, and complete practice questions to assist you in retaining the content.

6. DON'T GIVE UP

As you begin your journey towards nursing, keep in mind that the road will be tough and there will be moments where you may feel defeated; in the midst of these temporary moments, remember to push forward. There will be good days, there will be bad days, there will be days where studying will be the last thing on your mind; in the midst of these temporary moments, remember to push forward. You are in control of your destiny, and it is you who will decide how far you will go. No matter what life throws at you, always remember to never give up!

4-week schedule

	Monday	Tuesday	Wednesday	Thursday	Friday	Saturday	Sunday
Week 1	**Diagnostic** Cluster 1	Cluster 1	Cluster 1	Cluster 2	Cluster 2	Cluster 2	**Review 1-2**
Week 2	Cluster 3	Cluster 3	Cluster 3	Review	Review	Review	**Review 1-3**
Week 3	Cluster 4	Cluster 4	Cluster 5	Cluster 5	Cluster 5	Cluster 5	**Review 1-5**
Week 4	**Diagnostic Retest**	**Review**	**Review**	**Practice Test 1**	**Review**	**Review**	**Practice Test 2**

6-week schedule

	Monday	Tuesday	Wednesday	Thursday	Friday	Saturday	Sunday
Week 1	**Diagnostic** Cluster 1	Cluster 1	Cluster 1	Cluster 1	Cluster 1	Cluster 1	**Review**
Week 2	Cluster 2	Cluster 2	Cluster 2	Cluster 2	Cluster 2	**Review 1-2**	**Review 1-2**
Week 3	Cluster 3	Cluster 3	Cluster 3	Cluster 3	Cluster 3	**Review 1-3**	**Review 1-3**
Week 4	Cluster 4	Cluster 4	Cluster 4	Cluster 4	Cluster 4	**Review 1-4**	**Review 1-4**
Week 5	Cluster 5	Cluster 5	Cluster 5	Cluster 5	Cluster 5	**Review 1-5**	**Review 1-5**
Week 6	**Diagnostic Retest**	**Review**	**Review**	**Practice Test 1**	**Review**	**Review**	**Practice Test 2**

8-week schedule

	Monday	Tuesday	Wednesday	Thursday	Friday	Saturday	Sunday
Week 1	**Diagnostic**	Cluster 1	Cluster 1	Cluster 1	Cluster 1	Cluster 1	**Review**
Week 2	**Review**	Cluster 2	Cluster 2	Cluster 2	Cluster 2	Cluster 2	**Review**
Week 3	**Review**	Cluster 3	Cluster 3	Cluster 3	Cluster 3	Cluster 3	**Review**
Week 4	**Review**	Cluster 4	Cluster 4	Cluster 4	Cluster 4	Cluster 4	**Review**
Week 5	**Review**	Cluster 5	Cluster 5	Cluster 5	Cluster 5	Cluster 5	**Review**
Week 6	**Review**	**Diagnostic**	**Review**	**Review**	**Practice Test 1**	**Review**	**Review**
Week 7	**Practice Test 2**	**Review**	**Review**	**Review**	**Diagnostic**	**Review**	**Review**
Week 8	**Practice Test**	**Review**	**Review**	**Practice Test 2**	**Review**	**Review**	**Review**

K.I.S.S IT
TEAS Math Edition

Diagnostic Test

The diagnostic test was created in order to gauge your current knowledge level of the topics included on the test. The diagnostic test can be used as a guide to help you focus on specific topics within the book.

Clustered Topics

The math section of the TEAS test covers 15 topics, and for most this can be a little overwhelming. This workbook groups topics together based on similar properties they share in terms of sub-skills and methods of solving. Each skill was specifically placed in order in an effort to strengthen subskills and illustrate how many of the topics featured in the test are interrelated.

Limited Math Terminology

I know that most of you all are terrified of math which creates a high level of anxiety. All of the topics featured in this book are explained in laments terms in order for you to better retain the information. Your future career in nursing requires limited math terminology, so this serves as the perfect guide for you.

Over 1,400 Practice Questions

Perfect practice makes perfect! This book is equipped with 1,413 practice questions that will help you prepare for your upcoming exam.

2 Full-length Practice Tests

The practice tests featured in this book were created in order to mimic the actual test. Use these tools to gauge how well you have progressed, and which topics you need to review if necessary.

Table of Contents

Diagnostic Test

36 Questions - 54 minutes

1. Estimate $\sqrt{10}$ (to the nearest hundredth)

a. 3.16 b. 5 c. 3.17 d. 3.162

2. Which unit is best when estimating the weight of a small child?

a. Grams b. liters c. feet d. kilograms

3. Identify the fraction that is equivalent to 0.32

a. $\frac{32}{10}$ b. $\frac{8}{25}$ c. $\frac{32}{1000}$ d. $\frac{16}{25}$

4. Convert $\frac{7}{8}$ to a percentage

a. 78% b. 114% c. 87.5% d. 8.75%

5. Identify the largest value: 1.675 1.68 1.673 1.6

a. 1.675 b. 1.68 c. 1.673 d. 1.6

6. Arrange the values in ascending order: $-\frac{1}{4}$ $\sqrt{9}$ -2 160%

a. $\sqrt{9}$ 160% $-\frac{1}{4}$ -2
b. -2 $-\frac{1}{4}$ 160% $\sqrt{9}$
c. $\sqrt{9}$ 160% -2 $-\frac{1}{4}$
d. $-\frac{1}{4}$ -2 160% $\sqrt{9}$

7. 389 g = _________________ kg

a. 0.389 b. 3.89 c. 38.9 d. 3,890

8. 3.24 liters = _______________ quarts (1 quart = 0.95 liter)

a. 3.078 b. 3.411 c. 0.293 d. 3.029

9. A bag contains 32 marbles: 6 red, 15 black, 7 orange, and 4 green. What is the ratio of red and black marbles to the total number of marbles in the bag?

a. 3:16 b. 15:32 c. 21:32 d. 6:15

10. The ratio of men to women at the local factory is 5:1. A factory located 20 miles away has an equivalent ratio of men to women. Identify the ratio that represents the number of men to women at the neighboring factory.

a. 25:26 b. 1:5 c. 10:6 d. 15:3

11. The ratio of white to black socks in the top drawer is 1:3. If there are 11 white socks in the drawer, how many black socks are there?

a. 33 b. 13 c. 3.67 d. 30

12. Roman drives 250 miles on 28 gallons of gas. If he travels 775 miles, how many gallons of gas will he use?

a. 86.8 b. 69 c. 553 d. 85

13. Taylor uses $2\frac{1}{3}$ cups of sugar to bake a dozen cookies. How much sugar will she need to bake 48 cookies?

a. 112 b. 96 c. $9\frac{1}{3}$ d. 9

14. Last year, there were 150 students enrolled in the Law Academy at the local high school. This year, enrollment has increased to 225 students. What is the percentage increase in students enrolled?

a. 75% b. 50% c. 33% d. 60%

15. Tori purchased a pair of jeans for $24.99, a blouse for $12.95, and a necklace for $6.99. If the sales tax rate is 6.5% in Florida, how much will she pay **in taxes?**

a. $44.93 b. $47.85 c. $2.65 d. $2.92

16. The local high school basketball team won 32 out of 45 games this season. What percentage of their games did they lose?

a. 13% b. 71% c. 29% d. 87%

17. A school district studies the impact of teacher pay on job satisfaction. Identify the independent variable in the given scenario.

a. District b. Teacher Pay c. Satisfaction d. Impact

18. Identify the type of correlation illustrated in the graph below

a. positive correlation b. negative correlation c. no correlation

19. Based on the graph, what percentage of the scholarships awarded were given to baseball players?

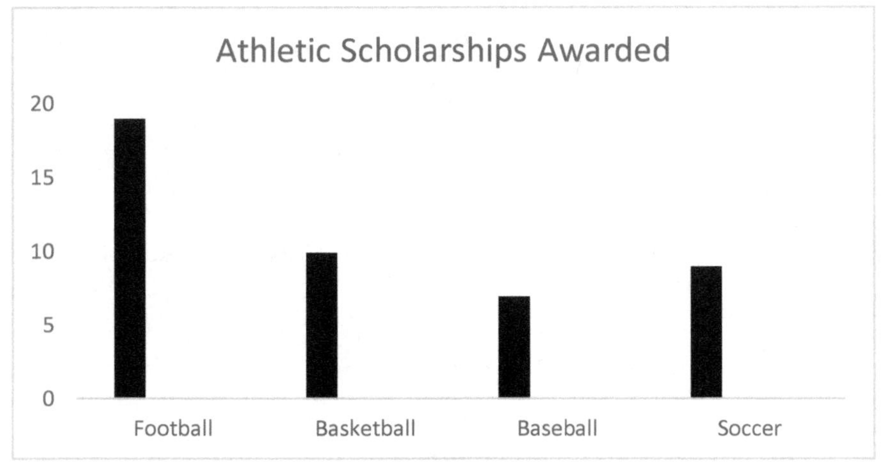

a. 10%　　b. 16%　　c. 20%　　d. 26%

20. Which graph would best illustrate a state's graduation rates from the last five years?

a. Histogram　　b. Pie Chart　　c. Scatter Plot　　d. Line Graph

21. Which graph shape has a mean that is less than the median?

a. skewed right　　b. skewed left　　c. uniform　　d. normal

22. Given the data set {2,6,8,4,6,10} calculate the mean.

a. 6　　b. 8　　c. 36　　d. 10

23. Anna's favorite box of cereal is larger than the average box with dimensions of 15 inches by 10 inches. What is the area of the box's face?

a. 50 in^2　　b. 150 in　　c. 150 in^2　　d. 50 in

24. Nick's pizza place is home to the city's largest pizza pie with an area of 85 square inches. What is the pie's radius? (round to the nearest hundredth)

a. 5.20　　b. 27.02　　c. 13.54　　d. 5.21

25. $2\frac{1}{8} \times 4\frac{2}{3}$

a. $8\frac{1}{12}$　　b. $9\frac{11}{12}$　　c. $\frac{12}{119}$　　d. $8\frac{3}{16}$

26. $4\frac{1}{9} \div 1\frac{2}{7}$

a. $4\frac{7}{18}$　　b. $3\frac{16}{81}$　　c. $5\frac{2}{7}$　　d. $3\frac{2}{7}$

27. $\frac{9}{8} - \frac{7}{11}$

a. $\frac{2}{3}$　　b. $\frac{16}{19}$　　c. $1\frac{67}{88}$　　d. $\frac{43}{88}$

28. $\frac{3}{5}+\frac{1}{3}+\frac{2}{4}$

a. $1\frac{13}{30}$ b. $\frac{1}{2}$ c. $1\frac{1}{10}$ d. $\frac{13}{30}$

29. Luke can type a 10-page paper in 12 hours, Cindy can type the same number of pages in 8 hours. If they work together, how long will it take them to complete the paper?

a. 9 hours b. .21 hours c. 5.5 hours d. 4.8 hours

30. Ashley is paid $2,754.89 bi-monthly. If $432.58 in taxes is taken with each paycheck, what is her net annual salary?

a. $55,735.44 b. $33,058.68 c. $27,867.72 d. $76,499.28

31. At the local supermarket, oranges are $1.29/lb., apples are 75 cents/lb., and bananas are 30 cents/lb. If Ana purchases 3 lbs. of oranges, 2 lbs. of apples, and 5 lbs. of bananas, what is her total?

a. $303.87 b. $2.34 c.$10 d. $6.87

32. Kendrick's checking account had an available balance of $20,235.54 on Wednesday. That same day, he made a purchase for $8,298.21. Thursday, his bank charged him a monthly fee of $25. He wrote a check the next day for $1000. What was his available balance Friday?

a. $27,508.75 b. $12,962.33 c. $10,912.33 d. $11,912.33

33. Seventeen more than the product of a and 7 is forty-two

a. $17a + 7 = 42$ b. $7a + 17$ c. $17a + 7$ d. $7a + 17 = 42$

34. five less than b is at most 23

a. $5 - b < 23$ b. $b - 5 \leq 23$ c. $b - 5 \geq 23$ d. $5 - b > 23$

35. The number of nursing majors is forty-two less than four times the number of communications majors. Express the number of nursing majors (n) in terms of the number of communication majors (c).

a. $c = 4n - 42$ b. $n = 42c - 4$ c. $n = 4c - 42$ d. $4c - 42$

36. One half of the sum of the number of apples and 6 equals eight. How many apples are there?

a. 4 b. 10 c. 6 d. 8

Answers & Explanations

1. A. $\sqrt{10} = 3.1622$ which rounds to 3.16 because the two to the right of 6 is less than 5 which means the 6 stays the same.

2. D. Grams are too small (a paperclip is about 1 gram), liters measure volume, and feet measure length. Kilograms is the best choice.

3. B. When converting a decimal to a fraction, place the number over its respective value. There are two numbers after the decimal, so in this case the denominator would be 100: $\frac{32}{100}$. The greatest common factor between 32 and 100 is 4, so the fraction reduces down to $\frac{8}{25}$.

4. C. When converting a fraction to a decimal, you divide the top number by the bottom number. $7 \div 8 = .875$. When converting a decimal to a percentage, you multiply by 100: $.875 \times 100 = 87.5\%$.

5. B. When comparing decimals with the same whole number, focus on the numbers after the decimal. The values include 1.**675** 1.**68** 1.**673** 1.**6.** Add zeros as placeholders, so that each value has the same number of digits. 1.**675** 1.**680** 1.**673** 1.**600.** The largest number is 680, so the answer is **1.68.**

6. B. After converting the values, you have: -.25 3 -2 1.60. Ascending means from least to greatest, so the negatives are the smallest values. -2 is smaller than -.25. Therefore, the correct order is -2, $-\frac{1}{4}$ 160% $\sqrt{9}$

7. A. When converting from grams to kilograms you need to move the decimal 3 spaces to the left. When you don't see the decimal, you automatically place it at the end of the number.

8. B. When converting from one unit to another, set up a ratio. 3.24 liters $\times \frac{\text{1 quart}}{\text{0.95 liter}}$. Because the two numbers are at different levels, you should divide thus giving you 3.411.

9. C. There are 21 red and black marbles (15+6), and there are 32 marbles in the bag. Therefore, the ratio is 21:32.

10. D. The ratio 5:1 is equivalent to 15:3. When you cross multiply the two ratios $\frac{5}{1} = \frac{15}{3}$, both sides are equal…15=15.

11. A. $\frac{\text{1 white}}{\text{3 black}} = \frac{\text{11 white}}{\text{? black}}$. After cross multiplying $(11 \times 3) \div 1 = 33$ black socks

12. A. $\frac{\text{250 miles}}{\text{28 gallons}} = \frac{\text{775 miles}}{\text{? gallons}}$. After cross multiplying $(775 \times 28) \div 250 = 86.8$ gallons

13. C. $\frac{\frac{7}{3}\,cups}{12\,cookies} = \frac{?\,cups}{48\,cookies}$. After cross multiplying $(\frac{7}{3} \times 48) \div 12 = 9\frac{1}{3}$

14. B. Percentage Increase $= \frac{225-150}{150} = 50\%$

15. D. Total purchase is \$44.93 (\$24.99 + \$12.95 + \$6.99). \$44.93 × .065 = \$2.92

16. C. The team lost 13 games (45 – 32). 13 losses ÷ 45 games total = 28.8% rounds to 29%

17. B. Teacher pay is not dependent on job satisfaction, job satisfaction is dependent on teacher pay; therefore, teacher pay is the independent variable.

18. B. The data is slanted downward from left to right, thus making it negative.

19. B. (19 + 10 + 7 + 9) = 45 total scholarships 7 baseball ÷ 45 total = 15.6 % ⟶ 16%

20. D. Line graph best represents this scenario because it's recording data over a period of time.

21. B. Graphs that are skewed left have a mean that is less than the median.

22. A. (2+6+8+4+6+10) = 36 ÷ 6 = 6.

23. C. Area = l × w = 15 × 10 = 150 in^2. The units are squared because it's the area.

24. A. Area = πr^2. $85 = 3.14r^2$ $85 \div 3.14 = r^2$. $27.07 = r^2$. $\sqrt{27.07} = r$. $5.20 = r$

25. B. $\frac{17}{8} \times \frac{14}{3} = \frac{238}{24} = 9\frac{11}{12}$

26. B. $\frac{37}{9} \div \frac{9}{7} \longrightarrow \frac{37}{9} \times \frac{7}{9} = \frac{259}{81} = 3\frac{16}{81}$

27. D. $\frac{9}{8} - \frac{7}{11} = \frac{99}{88} - \frac{56}{88} = \frac{43}{88}$

28. A. $\frac{3}{5} + \frac{1}{3} = \frac{5}{15} + \frac{9}{15} = \frac{14}{15} + \frac{2}{4} = \frac{30}{60} + \frac{56}{60} = \frac{86}{60} = 1\frac{13}{30}$

29. D. (12 × 8) ÷ (12 + 8) = 96 ÷ 20 = 4.8 hours

30. A. $2,754.89 - $432.50 = $2,322.31 each paycheck. $2,322.31 × 2 = $4,644.62 each month. $4,644.60 × 12 = $55,735.44 each year.

31. D. ($1.29 × 3) + (0.75 × 2) + (0.30 × 5) = $6.87

32. C. $20,235.54 - $8,298.21 - $25 - $1000 = $10,912.33

33. D. 17 more means add, the product of 7 and a means multiply, and is means equal.

34. B. Five less means subtract, and at most means less than or equal to.

35. C. Forty-two less than means subtract, 4 times the number of communication majors means multiply. Writing n in terms of c means that n = c.

36. B. $\frac{1}{2}(a + 6) = 8 \longrightarrow \frac{1}{2}a + 3 = 8 \longrightarrow a = 10$

Diagnostic Break Down

Apply estimation strategies...

Question 1 ☐

Question 2 ☐

Convert among non-negative fractions...

Question 3 ☐

Question 4 ☐

Compare and order rational numbers

Question 5 ☐

Question 6 ☐

Convert within and between standard...

Question 7 ☐

Question 8 ☐

Solve problems involving ratios...

Question 9 ☐

Question 10 ☐

Question 11 ☐

Solve problems involving proportions

Question 12 ☐

Question 13 ☐

Solve problems involving percentages

Question 14 ☐

Question 15 ☐

Question 16 ☐

Explain the relationship btw var...

Question 17 ☐

Question 18 ☐

Interpret relevant information from....

Question 19 ☐

Question 20 ☐

Evaluate the information in tables...

Question 21 ☐

Question 22 ☐

Calculate geometric quantities

Question 23 ☐

Question 24 ☐

Perform arithmetic operations...

Question 25 ☐

Question 26 ☐

Question 27 ☐

Question 28 ☐

Solve problems rational numbers

Question 29 ☐

Question 30 ☐

Question 31 ☐

Question 32 ☐

Translate phrases and sentences into...

Question 33 ☐

Question 34 ☐

Solve equations in one variable

Question 35 ☐

Question 36 ☐

HOW TO USE DIAGNOSTIC SCORE & CHECKLIST

Diagnostic Score

The diagnostic test is used as a tool to gauge your content knowledge prior to studying. The purpose of the diagnostic is to provide you with some insight as to which topics you've already mastered, and which ones you need assistance with. It is not uncommon for you to score poorly on this initial assessment, as the goal of this workbook is to assist you in strengthening your areas of weakness.

Diagnostic Checklist

The checklist is a tool that you can utilize in order to pinpoint the exact topic areas that you are weak in. Place an "X" in each box located next to the questions that you missed and highlight the topic areas. Use this score as a guide if you do not have enough time to complete the entire workbook.

Cluster 1

Apply Estimation Strategies and Rounding Rules to Real-World Problems

- How to round whole numbers and decimals
- How to simplify a radical
- How to choose an appropriate tool of measure
- How to estimate the mass, length, and width of everyday objects.

Convert Among Non-Negative Fractions, Decimals, and Percentages

- How to convert fractions to decimals
- How to convert fractions to percentages
- How to convert decimals to percentages
- How to convert decimals to fractions
- How to convert percentages to fractions
- How to convert percentages to decimals

Compare and Order Rational Numbers

- How to compare rational values
- How to identify the largest value in a data set
- How to identify the smallest value in a data set
- How to arrange values in ascending order
- How to arrange values in descending order

Convert Within and Between Standard and Metric Systems

- How to convert within metric and standard systems
- How to convert between metric and standard systems

Apply Estimation Strategies & Rounding Rules to Real-World Problems

You will learn:

How to round whole numbers and decimals

How to simplify a radical

How to choose an appropriate tool of measure

How to estimate the mass, length, and width of everyday objects.

Study Tips

Read and study EVERY example problem.

Complete EVERY practice problem.

Check to make sure all answers are correct.

Go back to correct the questions you answered incorrectly.

If you don't receive at least an 80% on the review, go back and practice the topic.

ROUNDING

KISS IT!

Step 1: Underline place value

Step 2: Assess whether the value stays the same or increases by 1

Hundred -Thousands	Ten -Thousands	Thousands	Hundreds	Tens	Ones		Tenths	Hundredths	Thousandths	Ten -Thousandths	Hundred -Thousandths
3	6	5	1	9	2	•	4	8	7	5	2

Example 1

Round 365,192.4875 to the nearest **hundredth**

Step 1: Underline place value

365,192.4<u>8</u>75

Step 2: Assess whether the value stays the same or increases by 1

- Look at the value to the right of the underlined number
- IF THE VALUE IS 4 OR LESS, THEN THE UNDERLINED NUMBER REMAINS THE SAME
- IF THE VALUE IS 5 OR MORE, THEN THE UNDERLINED NUMBER INCREASES BY 1

365,192.4<u>8</u>75

↓

ASSESS THE VALUE: 7 is greater than 5, so 8 will increase to 9. **Each value after the underlined number will change to zeros.**

365,192.490

Example 2

Round 365,192.4875 to the nearest **thousand**

Step 1: Underline place value

36<u>5</u>, 192. 4875

Step 2: Assess whether the value stays the same or increases by 1

- Look at the value to the right of the underlined number
- IF THE VALUE IS 4 OR LESS, THEN THE UNDERLINED NUMBER REMAINS THE SAME
- IF THE VALUE IS 5 OR MORE, THEN THE UNDERLINED NUMBER INCREASES BY 1

36<u>5</u>,192.4875

↓

ASSESS THE VALUE: 1 is less than 4, so 5 will remain the same. **Each value after the underlined number will change to zeros**

365,000.0000

Practice 1: Round the values to the given place value

Nearest Hundred		Nearest Tenth		Nearest Ten	
1. 3,456	**6.** 320	**11.** 345.789	**16.** 78.983	**21.** 3,456	**26.** 328
2. 279	**7.** 5,247	**12.** 24.649	**17.** 62.17	**22.** 279	**27.** 194
3. 134,789	**8.** 76	**13.** 3.54	**18.** 32.08	**23.** 134,789	**28.** 3,645
4. 949	**9.** 23,499	**14.** 12.765	**19.** 21.80	**24.** 949	**29.** 12.032
5. 13,589	**10.** 106	**15.** 435.923	**20.** 132.49	**25.** 13,589	**30.** 12,098

Nearest Thousandth		Nearest Thousand		Nearest Hundredth	
31. 345.7894	**36.** 0.2765	**41.** 3,456	**46.** 745	**51.** 345.789	**56.** 103.894
32. 24.6489	**37.** 2.9458	**42.** 2,793	**47.** 12,499	**52.** 24.469	**57.** 2.326
33. 3.5472	**38.** 0.6832	**43.** 134,789	**48.** 324,975	**53.** 3.542	**58.** 89.541
34. 12.7657	**39.** 34.1428	**44.** 9,498	**49.** 1,299	**54.** 12.765	**59.** 0.745
35. 435.9236	**40.** 22.4324	**45.** 13,589	**50.** 45,932	**55.** 435.923	**60.** 9.054

SIMPLIFYING RADICALS

KISS IT!

Step 1: Utilize Your Calculator!

Example 3

Simplify $\sqrt{7}$. Round to the nearest **tenth**

Step 1: Tap 7 on your calculator, then hit the $\sqrt{\ }$ symbol

$\sqrt{7} = 2.6458 \longrightarrow 2.6$

Example 4

Simplify $\sqrt{15}$. Round to the nearest **hundredth**

Step 1: Tap 15 on your calculator, then hit the $\sqrt{\ }$ symbol

$\sqrt{15} = 3.8729 \longrightarrow 3.87$

Example 5

Simplify $\sqrt{7}$ Round to the nearest **thousandth**

Step 1: Tap 7 on your calculator, then hit the $\sqrt{\ }$ symbol

$\sqrt{7} = 2.6458 \longrightarrow 2.646$

Practice 2 (Round your answer to the nearest hundredth)

1. $\sqrt{90}$ = ___________

2. $\sqrt{245}$ = ___________

3. $\sqrt{14}$ = ___________

4. $\sqrt{20}$ = ___________

5. $\sqrt{145}$ = ___________

6. $\sqrt{94}$ = ___________

7. $\sqrt{48}$ = ___________

8. $\sqrt{75}$ = ___________

9. $\sqrt{82}$ = ___________

10. $\sqrt{113}$ = ___________

11. $\sqrt{34}$ = ___________

12. $\sqrt{7}$ = ___________

13. $\sqrt{32}$ = ___________

14. $\sqrt{56}$ = ___________

15. $\sqrt{62}$ = ___________

16. $\sqrt{123}$ = ___________

17. $\sqrt{436}$ = ___________

18. $\sqrt{365}$ = ___________

19. $\sqrt{287}$ = ___________

20. $\sqrt{105}$ = ___________

21. $\sqrt{23}$ = ___________

22. $\sqrt{156}$ = ___________

23. $\sqrt{176}$= ___________

24. $\sqrt{89}$ = ___________

25. $\sqrt{234}$= ___________

26. $\sqrt{389}$ = ___________

27. $\sqrt{657}$= ___________

28. $\sqrt{275}$= ___________

29. $\sqrt{79}$= ___________

30. $\sqrt{92}$= ___________

CHOOSING APPROPRIATE TOOLS OF MEASURE

Units of Measure

Length	Volume	Weight/Mass
Inches Feet (ruler = 1 foot) Yards (3 feet = 1 yd) Meters Miles	Liter Gallon Quart Pint	Grams Ounces Pounds (16 oz) Tons (2,000 lbs.)

Measuring Tools

Length	Volume	Weight/Mass
Ruler (1 foot) Meter Stick Caliper Yard Stick (3 rulers) Measuring Tape	Pipette Burette Graduated Cylinder Flask Beaker	Scale Balance Measuring cup Measuring spoon

KISS IT!

Step 1: Identify the unit of measure

Step 2: Eliminate the unit of measure that doesn't fit

Step 3: Choose the most appropriate unit of measure

Example 6

What's the best unit of measure for the mass of a cell phone?

a. Tons b. Inches c. Pint d. Ounces

Step 1: Identify the unit of measure

What's the best unit of measure for the **mass** of a cell phone?

Step 2: Eliminate the unit of measure that doesn't fit

a. Tons b. ~~Inches~~ c. ~~Pint~~ d. Ounces

Length **Volume**

Step 3: Choose the most appropriate unit of measure

a. Tons b. ~~Inches~~ c. ~~Pint~~ d. Ounces

Too large **Most appropriate**

a. Tons b. Inches c. Pint **d. Ounces**

Practice 3

1. What's the best unit of measure for the weight of a dirt bike?

a. meters b. pounds c. tons d. gallons

2. What's the best unit of measure for the height of coffee table?

a. pounds b. yards c. inches d. meters

3. What's the best unit of measure for the width of a notebook?

a. inches b. centimeters c. pounds d. liters

4. What's the best unit of measure for the amount of water in a small glass?

a. pounds b. inches c. cups d. gallons

5. What's the best unit of measure for the weight of a towel?

a. grams b. meters c. tons d. pounds

6. What's the best unit of measure for the amount of water in a large swimming pool?

a. gallons b. ounces c. grams d. liters

7. What's the best unit of measure for the length of a thumb?

a. meters b. centimeters c. inches d. ounces

8. What's the best unit of measure for the mass of a television remote?

a. meters b. grams c. pounds d. liters

9. What's the best unit of measure for the mass of a leaf?

a. grams b. millimeters c. tons d. liters

10. What's the best unit of measure for the thickness of an ice-cream sandwich?

a. inches b. meters c. gallons d. liters

11. What's the best unit of measure for the mass of a car?

a. pounds b. ounces c. liters d. tons

12. What's the best unit of measure for the width of a standard refrigerator?

a. pounds b. inches c. feet d. ounces

13. What's the best unit of measure for the amount of liquid an eye dropper can hold?

a. cups b. liters c. milliliters d. feet

14. What's the best unit of measure for the mass of a small tire?

a. pounds b. tons c. meters d. liters

Example 7

Estimate the height of a standard refrigerator

a. 1.75 feet b. 1.75 yards c. 1.75 kilometers d. 1.75 centimeters

Step 1: Identify the units of measure

a. 1.75 **feet** b. 1.75 **yards** c. 1.75 **kilometers** d. 1.75 **centimeters**

Step 2: Identify items that are SIMILAR in length to the units

a. 1.75 **feet** → **RULER**

b. 1.75 **yards** → **3 RULERS**

c. 1.75 **kilometers** → **CLOSE TO A MILE**

d. 1.75 **centimeters** → **WIDTH OF THUMB**

Step 3: Choose the most appropriate unit of measure

a. 1.75 **feet** → **too small**

b. 1.75 **yards** → **most appropriate**

c. 1.75 **kilometers** → **too large**

d. 1.75 **centimeters** → **too small**

a. 1.75 feet (b. 1.75 yards) c. 1.75 kilometers d. 1.75 centimeters

Practice 4

1. Estimate the weight of a small sofa.

a. 75 oz b. 75 grams c. 75 pounds d. 75 tons

2. Estimate the height of a standard door.

a. 80 centimeters b. 80 millimeters c. 80 inches d. 80 feet

3. Estimate the length of a bottle of lotion.

a. 8 inches b. 8 centimeters c. 8 yards d. 8 millimeters

4. Estimate the mass of an eraser cap.

a. 3 milligrams b. 3 grams c. 3 kilograms d. 3 centigrams

5. Estimate the volume of a small glass of milk.

a. 8 milliliters b. 8 liters c. 8 ounces d. 8 grams

6. Estimate the mass of a small dog.

a. 0.2 grams b. 0.2 kilograms c. 0.2 milligrams d. 0.2 centigrams

7. Estimate the width of a small laptop.

a. 30 centimeters b. 30 inches c. 1 meter d. 1 yard

8. Estimate the mass of a standard refrigerator.

a. 300 ounces b. 300 tons c. 300 pounds d. 300 grams

9. Estimate the width of a thumb.

a. 1inch b. 1 centimeter c. 1 millimeter d. 1 meter

10. Estimate the mass of a small laptop.

a. 5 ounces b. 5 pounds c. 5 grams d. 5 liters

11. Estimate the volume of water in a standard swimming pool.

a. 13,000 ounces b. 13,000 gallons c. 13,000 pounds d. 13,000 liters

12. Estimate the length of standard football field.

a. 120 yards b. 120 feet c. 120 kilometers d. 120 miles

13. Estimate the mass of an apple.

a. 100 pounds b. 100 ounces c. 100 grams d. 100 kilograms

14. Estimate the length of your forearm.

a. 1.5 feet b. 1.5 centimeters c. 1.5 yards d. 1.5 meters

15. Estimate the length one eyelash.

a. 10 inches b. 10 centimeters c. 10 millimeters d. 10 meters

16. Estimate the volume of liquid in a small soda can.

a. 8 liters b. 8 ounces c. 8 pints d. 8 quarts

17. Estimate the mass of a standard smart phone.

a. 175 ounces b. 175 pounds c. 175 kilograms d. 175 grams

18. Estimate the mass of a newborn child.

a. 100 ounces b. 100 grams c. 100 pounds d. 100 liters

19. Estimate the length of your ring finger

a. 3 centimeters b. 3 meters c. 3 feet d. 3 inches

20. Estimate the mass of a cow.

a. 1,600 tons b. 1,600 ounces c. 1,600 pounds d. 1,600 grams

21. Estimate the mass of a full-sized mattress.

a. 50 pounds b. 50 ounces c. 50 grams d. 50 inches

Review

1. Round 12.345 to the nearest hundredth.

2. What is the best unit of measure for the width of a standard flash card?

a. kilometers b. meters c. inches d. liters

3. Estimate the sum of 34,654 + 45,923 to the nearest ten thousand.

4. $\sqrt{52}$ (nearest thousandth) =

5. Estimate the weight of a small laptop.

a. 5 mg b. 5 g c. 2 kg d. 2 cg

6. What is the best unit of measure for the length of a standard football field?

a. miles b. feet c. yards d. inches

7. Round 8.6784 to the nearest whole number.

8. What is the best unit of measure for the thickness of a fingernail?

a. milliliter b. centiliter c. millimeter d. centimeter

9. Estimate the product of 356 × 420 to the nearest hundred.

10. $\sqrt{104}$ (nearest hundredth) =

11. Round 123,456.789498 to the nearest thousandth.

12. Round 456.92 to the nearest tenth.

13. $\sqrt{67}$ (nearest whole number) =

14. $\sqrt{.009}$ (nearest hundredth) =

15. Round 135,495.45 to the nearest thousand.

16. Estimate the length of a twin sized mattress.

a. 75 centimeters b. 75 inches c. 75 feet d. 75 meters

17. $\sqrt{1.21}$ (nearest hundredth) =

18. Estimate the height of the Empire State Building.

a. 1400 miles b. 1400 feet c. 1400 yards d. 1400 millimeters

19. $\sqrt{3}$ (nearest tenth) =

20. Estimate the mass of a small Pandora charm.

a. 20 milligrams b. 20 centigrams c. 20 grams d. 20 kilograms

Answer Key

Practice 1

1. 3500	**11.** 345.8	**21.** 3,460	**31.** 345.789	**41.** 3,000	**51.** 345.79
2. 300	**12.** 24.6	**22.** 280	**32.** 24.649	**42.** 3,000	**52.** 24.47
3. 134,800	**13.** 3.5	**23.** 134,790	**33.** 3.547	**43.** 135,000	**53.** 3.54
4. 900	**14.** 12.8	**24.** 950	**34.** 12.766	**44.** 9,000	**54.** 12.77
5. 13,600	**15.** 435.9	**25.** 13,590	**35.** 435.924	**45.** 14,000	**55.** 435.92
6. 300	**16.** 79.0	**26.** 330	**36.** 0.277	**46.** 1,000	**56.** 103.89
7. 5,200	**17.** 62.2	**27.** 190	**37.** 2.946	**47.** 12,000	**57.** 2.33
8. 100	**18.** 32.1	**28.** 3,650	**38.** 0.683	**48.** 325,000	**58.** 89.54
9. 23,500	**19.** 21.8	**29.** 10	**39.** 34.143	**49.** 1,000	**59.** 0.75
10. 100	**20.** 132.50	**30.** 12,100	**40.** 22.432	**50.** 46,000	**60.** 9.05

Practice 2

1. 9.49	**6.** 9.70	**11.** 5.83	**16.** 11.09	**21.** 4.80	**26.** 19.72
2. 15.65	**7.** 6.93	**12.** 2.65	**17.** 20.88	**22.** 12.49	**27.** 25.63
3. 3.74	**8.** 8.66	**13.** 5.66	**18.** 19.10	**23.** 13.27	**28.** 16.58
4. 4.47	**9.** 9.06	**14.** 7.48	**19.** 16.94	**24.** 9.43	**29.** 8.89
5. 12.04	**10.** 10.63	**15.** 7.87	**20.** 10.25	**25.** 15.30	**30.** 9.59

Practice 3

1. B	**4.** C	**7.** C	**10.** A	**13.** C
2. C	**5.** D	**8.** C	**11.** D	**14.** A
3. A	**6.** A	**9.** A	**12.** B	

Practice 4

1. C	**5.** C	**9.** A	**13.** C	**17.** D	**21.** A
2. C	**6.** B	**10.** B	**14.** A	**18.** A	
3. A	**7.** A	**11.** B	**15.** C	**19.** D	
4. B	**8.** C	**12.** A	**16.** B	**20.** C	

Review

1. 12.35	**6.** C	**11.** 123,456.789	**16.** B
2. C	**7.** 9	**12.** 456.9	**17.** 1.10
3. 80,000	**8.** C	**13.** 8	**18.** B
4. 7.211	**9.** 160,000	**14.** 0.09	**19.** 1.7
5. C	**10.** 10.20	**15.** 135,000	**20.** C

Convert Among Non-Negative Fractions Decimals and Percentages

You will learn:

How to convert fractions to decimals

How to convert fractions to percentages

How to convert decimals to percentages

How to convert decimals to fractions

How to convert percentages to fractions

How to convert percentages to decimals

Study Tips

Read and study EVERY example problem.

Complete EVERY practice problem.

Check to make sure all answers are correct.

Go back to correct the questions you answered incorrectly.

If you don't receive at least an 80% on the review, go back and review the topic.

Fractions to Decimals (*HOT TOPIC)

KISS IT!

Step 1: Divide the top number by the bottom number

Example 1

Convert $\frac{1}{8}$ to a decimal

$1 \div 8 = 0.125$

$\frac{1}{8} = 0.125$

Example 2

Convert $2\frac{1}{4}$ to a decimal

$1 \div 4 = 0.25$

$2\frac{1}{4} = 2.25$

When you're given a mixed number, focus on the fraction when converting. After the conversion, place the whole number to the left of the decimal.

Practice 1: Convert the fractions to decimals (round to nearest thousandth)

1. $\frac{2}{9}$	**4.** $\frac{25}{100}$	**7.** $\frac{30}{100}$	**10.** $\frac{3}{8}$	**13.** $2\frac{32}{80}$	**16.** $\frac{13}{8}$
2. $\frac{1}{2}$	**5.** $2\frac{1}{8}$	**8.** $4\frac{3}{8}$	**11.** $\frac{7}{11}$	**14.** $\frac{11}{13}$	**17.** $\frac{2}{15}$
3. $\frac{2}{3}$	**6.** $\frac{9}{2}$	**9.** $9\frac{20}{100}$	**12.** $\frac{13}{5}$	**15.** $5\frac{9}{13}$	**18.** $\frac{1}{12}$

Fractions to Percentages

KISS IT!

Step 1: Divide the top number by the bottom number (Convert to decimal)

Step 2: Move the decimal two places to the right

Example 3

Convert $\frac{3}{8}$ to a percent

$3 \div 8 = 0.375$

$0.375 \times 100 = 37.5$

$\frac{3}{8} = $ **37.5%**

Example 4

Convert $2\frac{2}{9}$ to a percent

$2 \div 9 = 0.22$

$2.22 \times 100 = 222$

$2\frac{2}{9} = $ **222%**

When you're given a mixed number, focus on the fraction when converting. After the conversion, place the whole number to the left of the decimal.

Practice 2: Convert Fractions to Percentages (round to nearest tenth)

1. $\frac{2}{9}$ **4.** $\frac{25}{100}$ **7.** $\frac{20}{100}$ **10.** $\frac{3}{8}$ **13.** $2\frac{32}{80}$ **16.** $\frac{13}{8}$

2. $\frac{1}{2}$ **5.** $3\frac{2}{8}$ **8.** $4\frac{1}{9}$ **11.** $\frac{7}{11}$ **14.** $\frac{11}{13}$ **17.** $\frac{2}{15}$

3. $\frac{2}{3}$ **6.** $\frac{9}{2}$ **9.** $5\frac{2}{9}$ **12.** $\frac{13}{5}$ **15.** $5\frac{9}{13}$ **18.** $\frac{1}{12}$

Decimals to Percentages

KISS IT!

Step 1: Multiply decimal by 100

OR

Step 1: Move decimal two places to the right

Example 5

Convert 0.34 to a percentage

$0.34 \times 100 = 34$

0.34 = **34%**

OR

Convert 0.34 to a percentage

0.34.

0.34 = **34%**

Example 6

Convert 2.98 to a percentage

$2.98 \times 100 = 298$

2.98 = **298** %

OR

Convert 2.98 to a percentage

2.98.

2.98 = **298%**

Practice 3: Convert Decimals to Percentages

1. 0.45 **4.** 2.67 **7.** .0078 **10.** 0.345 **13.** 23.48 **16.** 0.8

2. 89.3 **5.** 0.9 **8.** 0.030 **11.** 2.78 **14.** 0.0048 **17.** 0.0034

3. .0045 **6.** 321.9 **9.** 0.07 **12.** 0.92 **15.** 0.034 **18.** 0.02

Mixed Review 1

1. .45 = % **6.** $\frac{40}{100}$ = decimal **11.** $\frac{9}{10}$ = decimal **16.** $3\frac{2}{5}$ = %

2. $\frac{4}{16}$ = % **7.** $2\frac{4}{8}$ = decimal **12.** 24.0 = % **17.** 0.8 = %

3. $\frac{3}{8}$ = decimal **8.** .0045 = % **13.** 0.0034 = % **18.** 2.98 = %

4. 6.23 =% **9.** $\frac{90}{8}$ = % **14.** $\frac{1}{2}$ = decimal **19.** $\frac{3}{5}$ = decimal

5. $\frac{25}{75}$ = % **10.** $\frac{7}{7}$ = % **15.** 12.5 = % **20.** $\frac{2}{9}$ = %

Decimal to Fraction

KISS IT!

Step 1: Use place value to determine the denominator

One number AFTER the decimal gives a denominator of **10**
Two numbers AFTER the decimal gives a denominator of **100**
Three numbers AFTER the decimal gives a denominator of **1000**
OR
Step 1: Use vocabulary to determine the denominator

(tenths, hundredths, thousandths)

Example 7

Convert 0.4 to a fraction **OR** Convert 0.4 to a fraction

How many numbers are AFTER the decimal? → 1

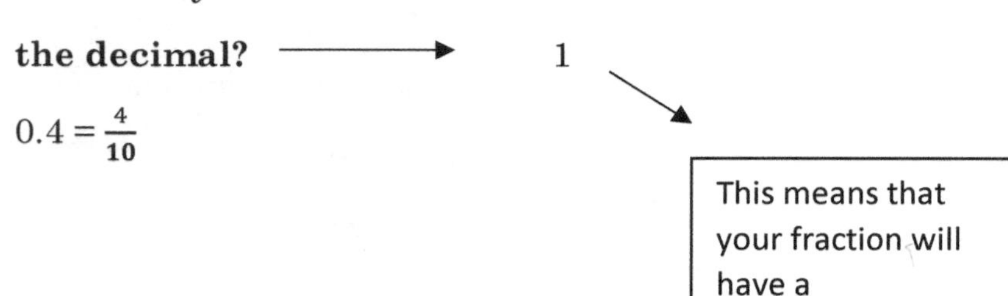

$0.4 = \frac{4}{10}$

This means that your fraction will have a **denominator of 10**

How do you say the number?

Four **tenths**

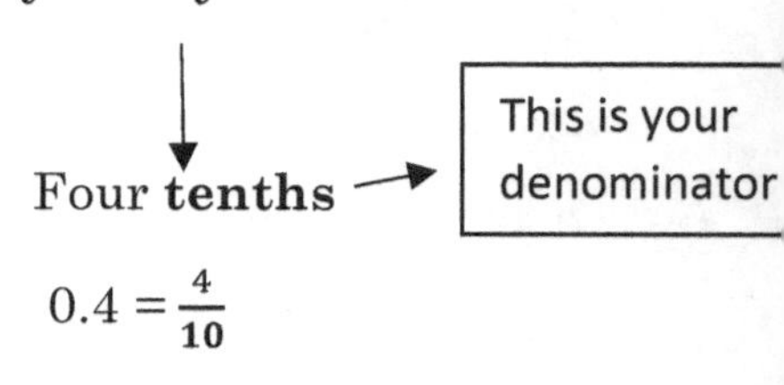

This is your denominator

$0.4 = \frac{4}{10}$

***DON'T FORGET TO REDUCE YOUR FRACTION**

$$\frac{4 \div 2}{10 \div 2} = \frac{2}{5}$$

Example 8

Convert 1.25 to a fraction **OR** Convert 1.25 to a fraction

How many numbers are AFTER the decimal? → 2

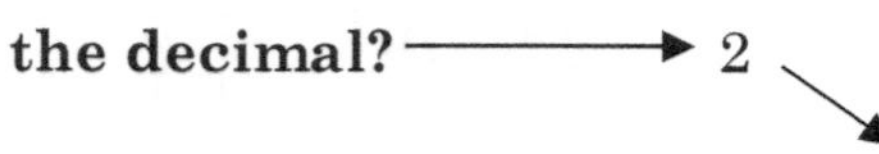

$1.25 = 1\frac{25}{100}$

This means that your fraction will have a **denominator of 100**

How do you say the number?

One and twenty-five **hundredths**

This is your denominator

$1.25 = 1\frac{25}{100}$

***DON'T FORGET TO REDUCE YOUR FRACTION**

$$1\frac{25}{100} = 1\frac{1}{4}$$

Example 9

Convert 0.045 to a fraction **OR** Convert 0. 045 to a fraction

How many numbers are AFTER the decimal? → **3**

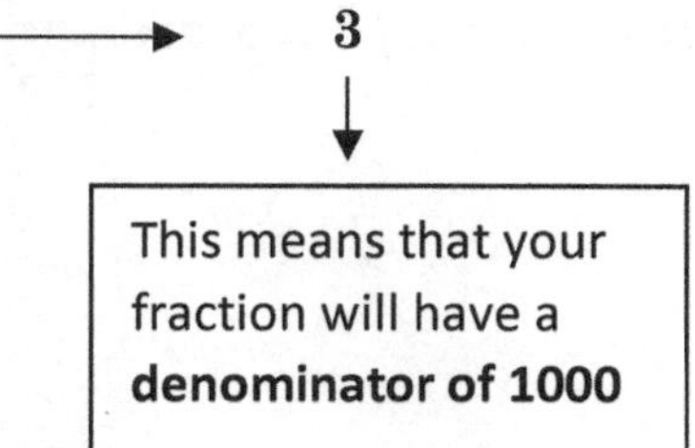

$0.045 = \frac{45}{1000}$

How do you say the number?

Forty-five **thousandths**

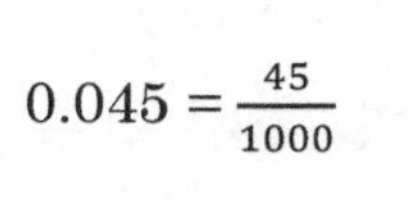

This is your denominator

***DON'T FORGET TO REDUCE YOUR FRACTION**

$$\frac{45 \div 5}{1000 \div 5} = \frac{9}{200}$$

Practice 4: Convert Decimals to Fractions

1. 1.06	**4.** 3.75	**7.** 23.375	**10.** 0.07	**13.** 0.65	**16.** 4.12
2. 0.125	**5.** 0.75	**8.** 0.006	**11.** 0.2	**14.** 1.8	**17.** 0.003
3. 0.4	**6.** 4.10	**9.** 2.6	**12.** 2.10	**15.** 0.32	**18.** 0.98

<u>Percent to Decimal</u>

KISS IT!

Step 1: Divide percentage by 100

OR

Step 1: Move decimal two places to the left

Example 10

Convert 25% to a decimal

$25 \div 100 = 0.25$

25% = 0.25

OR

.25.% = 0.25

Example 11

Convert 105% to a decimal

$105 \div 100 = 1.05$

105% = 1.05

OR

1.05.% = 1.05

Practice 5: Convert Percentages to Decimals

1. 205%	**4.** 5%	**7.** 148%	**10.** 3%	**13.** 239%	**16.** 7%
2. 36.25 %	**5.** .05%	**8.** 58.9%	**11.** 45%	**14.** .72%	**17.** 10%
3. 1.006%	**6.** 2.3%	**9.** .4%	**12.** 2.89%	**15.** 326%	**18.** 4%

Percent to Fraction

KISS IT!

Step 1: Convert to decimal

Step 2: Convert decimal to fraction

Example 12

Convert 30% to a fraction

Step 1: Convert 30% to a decimal

$30 \div 100 = 0.30$

Step 2: Convert decimal to fraction

$0.30 = \frac{30}{100} = \frac{3}{10}$

Example 13

Convert 5.6% to a fraction

Step 1: Convert 5.6% to a decimal

$5.6 \div 100 = .056$

Step 2: Convert decimal to fraction

$.056 = \frac{56}{1000} = \frac{7}{125}$

Practice 6: Convert Percentages to Fractions

1. 205%	**4.** 5%	**7.** 148%	**10.** 3%	**13.** 239%	**16.** 7%
2. 36%	**5.** 15%	**8.** 58%	**11.** 45%	**14.** .72%	**17.** 10%
3. 100%	**6.** 2.3%	**9.** 4%	**12.** 2.89%	**15.** 326%	**18.** 2%

Mixed Review 2

1. .04 = Frac	**6.** 0.3 = Frac	**11.** 183% = Dec	**16.** .2 = Frac
2. 150% = Frac	**7.** 1.8 = Frac	**12.** 8% = Frac	**17.** 1.34 = Frac
3. 69% = Dec	**8.** .04% = Dec	**13.** .37 = Frac	**18.** 23% = Dec
4. 4.5% = Dec	**9.** 10.5% = Frac	**14.** 4% = Dec	**19.** 789% = Dec
5. 75% = Frac	**10.** .255 = Frac	**15.** 82% = Frac	**20.** 76% = Frac

Answer Key

Practice 1	Practice 2	Practice 3	Mixed Review
1. .222	**1.** 22.2%	**1.** 45%	**1.** 45%
2. .500	**2.** 50%	**2.** 8930%	**2.** 25%
3. .667	**3.** 66.7%	**3.** .45%	**3.** .375
4. .250	**4.** 25%	**4.** 267%	**4.** 623%
5. 2.125	**5.** 325%	**5.** 90%	**5.** 33%
6. 4.500	**6.** 450%	**6.** 32190%	**6.** .40
7. .300	**7.** 20%	**7.** .78%	**7.** 2.50
8. 4.375	**8.** 411.1%	**8.** 3%	**8.** .45%
9. 9.200	**9.** 522.2%	**9.** 7%	**9.** 1125%
10. .375	**10.** 37.5%	**10.** 34.5%	**10.** 100%
11. .636	**11.** 63.6%	**11.** 278%	**11.** .90
12. 2.600	**12.** 260%	**12.** 92%	**12.** 2400%
13. 2.400	**13.** 240%	**13.** 2,348%	**13.** 0.34%
14. .846	**14.** 84.6%	**14.** 0.48%	**14.** 0.50
15. 5.692	**15.** 569.2%	**15.** 3.4%	**15.** 1,250%
16. 1.625	**16.** 162.5%	**16.** 80%	**16.** 340%
17. .133	**17.** 13.3%	**17.** 0.34%	**17.** 80%
18. .083	**18.** 8.3%	**18.** 2%	**18.** 298%
			19. .6
			20. 22.2%

Practice 4

1. $1\frac{6}{100} = 1\frac{3}{50}$

2. $\frac{125}{1000} = \frac{1}{8}$

3. $\frac{4}{10} = \frac{2}{5}$

4. $3\frac{75}{100} = 3\frac{3}{4}$

5. $\frac{75}{100} = \frac{3}{4}$

6. $4\frac{10}{100} = 4\frac{1}{10}$

7. $23\frac{375}{1000} = 23\frac{3}{8}$

8. $\frac{6}{1000} = \frac{3}{500}$

9. $2\frac{6}{10} = 2\frac{3}{5}$

10. $\frac{7}{100}$

11. $\frac{2}{10} = \frac{1}{5}$

12. $2\frac{10}{100} = 2\frac{1}{10}$

13. $\frac{65}{100} = \frac{13}{20}$

14. $1\frac{8}{10} = 1\frac{4}{5}$

15. $\frac{32}{100} = \frac{8}{25}$

16. $4\frac{12}{100} = 4\frac{3}{25}$

17. $\frac{3}{1000}$

18. $\frac{98}{100} = \frac{49}{50}$

Practice 5

1. 2.05

2. .3625

3. .01006

4. .05

5. .0005

6. .023

7. 1.48

8. .589

9. .004

10. .03

11. .45

12. .0289

13. 2.39

14. .0072

15. 3.26

16. .07

17. .10

18. .04

Practice 6

1. $2\frac{5}{100} = 2\frac{1}{20}$

2. $\frac{36}{100} = \frac{9}{25}$

3. $\frac{100}{100} = 1$

4. $\frac{5}{100} = \frac{1}{20}$

5. $\frac{15}{100} = \frac{3}{20}$

6. $\frac{2.3}{100} = \frac{23}{1000}$

7. $1\frac{48}{100} = 1\frac{12}{25}$

8. $\frac{58}{100} = \frac{29}{50}$

9. $\frac{4}{100} = \frac{1}{25}$

10. $\frac{3}{100}$

11. $\frac{45}{100} = \frac{9}{20}$

12. $\frac{289}{10{,}000}$

13. $2\frac{39}{100}$

14. $\frac{72}{10{,}000} = \frac{9}{1{,}250}$

15. $3\frac{26}{100} = 3\frac{13}{50}$

16. $\frac{7}{100}$

17. $\frac{10}{100} = \frac{1}{10}$

18. $\frac{2}{100} = \frac{1}{50}$

Mixed Review 2

1. $\frac{1}{25}$

2. $1\frac{50}{100} = 1\frac{1}{2}$

3. .69

4. .045

5. $\frac{75}{100} = \frac{3}{4}$

6. $\frac{3}{10}$

7. $1\frac{8}{10} = 1\frac{4}{5}$

8. .0004

9. $\frac{105}{1000} = \frac{21}{200}$

10. $\frac{255}{1000} = \frac{51}{200}$

11. 1.83

12. $\frac{8}{100} = \frac{2}{25}$

13. $\frac{37}{100}$

14. .04

15. $\frac{82}{100} = \frac{41}{50}$

16. $\frac{2}{10} = \frac{1}{5}$

17. $1\frac{34}{100} = 1\frac{17}{50}$

18. .23

19. 7.89

20. $\frac{76}{100} = \frac{19}{25}$

Compare and Order Rational Numbers

You will learn:

How to compare rational values

How to identify the largest value in a data set

How to identify the smallest value in a data set

How to arrange values in ascending order

How to arrange values in descending order

Study Tips

Read and study EVERY example problem.

Complete EVERY practice problem.

Check to make sure all answers are correct.

Go back to correct the questions you answered incorrectly.

If you don't receive at least an 80% on the review, go back and review the topic.

KISS IT!

Step 1: Convert the values to decimals

(UTILIZE YOUR CALCULATOR)

Sub-skills:

Convert fractions, decimals & percentages

Comparing decimals

Comparing positive & negative integers

Example 1

Identify the **largest** value

2.356 **2.355** $\frac{7}{3}$ $\frac{16}{6}$

Convert the fractions to decimals

(divide the top number by the bottom number)

$\frac{7}{3} = \mathbf{2.333}$

$\frac{16}{6} = \mathbf{2.667}$

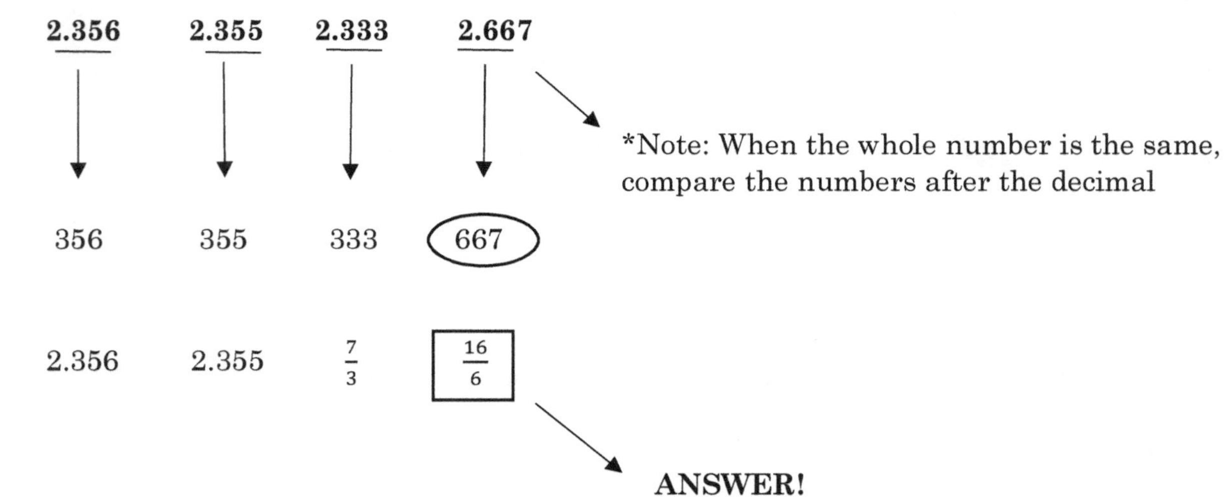

ANSWER!

Practice 1: Identify the largest value

1. 1.728 $\sqrt{3}$ $\frac{9}{5}$ 1.734
2. 68% $\sqrt{.25}$ $\frac{2}{3}$ .60
3. 2.664 2.654 2.689 2.688
4. -3 -4 0 -2 $\frac{1}{2}$
5. $-\sqrt{4}$ $\frac{3}{2}$ $\sqrt{4}$ 152%
6. -5 -3 -0.4 $-\frac{1}{2}$
7. -1.8667 -1.8689 -1.8668 -1.8660
8. 35% .345 $\frac{1}{3}$ $\sqrt{7}$
9. 6.545 6.544 6.546 6.549
10. .285 29% 28% .287

Practice 2: Identify the smallest value

1. -14% .03% .006 .01
2. -16 -42 -8 -35
3. $\sqrt{14}$ 3.73 378% $\frac{11}{3}$
4. $\frac{1}{3}$ $\frac{1}{2}$ $\frac{1}{8}$ $\frac{1}{4}$
5. 2.060 2.059 2.606 2.509
6. -3 -.45 -2 -.453
7. 25% $\frac{1}{5}$ .30 .301
8. $\frac{2}{3}$ $\frac{4}{6}$ $\frac{1}{2}$ $\frac{2}{6}$
9. -.65 -.649 -.651 -.64993
10. $\sqrt{25}$ 5.01 $\frac{26}{5}$ $\sqrt{26}$

Example 2

Arrange the values in **ascending** order (**ascending: least to greatest)**

$\frac{1}{2}$ 68% 0.51 $\sqrt{2}$ → Convert each value to a decimal

*Reference sub-skill: convert fractions, decimals & percentages

$\frac{1}{2}$ = .5 68% = .68 0.51 $\sqrt{2}$ = 1.41

0.5 0.68 0.51 1.41 → **Add zeros as place holders**

REARRANGE IN ASCENDING ORDER

0.5 0.51 0.68 1.41

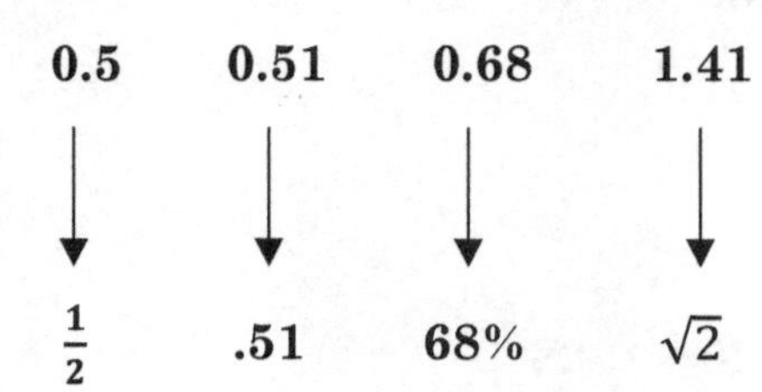

***Convert the decimals back to their original forms**

$\frac{1}{2}$ .51 68% $\sqrt{2}$

Practice 3: Arrange the values in ascending order

1. 1.728 $\sqrt{3}$ $\frac{9}{5}$ 1.734

2. 68% $\sqrt{.25}$ $\frac{2}{3}$ .60

3. 2.664 2.654 2.689 2.688

4. -3 -4 0 -2 $\frac{1}{2}$

5. $-\sqrt{4}$ $\frac{3}{2}$ $\sqrt{4}$ 152%

6. -5 -3 -0.4 $-\frac{1}{2}$

7. -1.8667 -1.8689 -1.8668 -1.8660

8. 35% .345 $\frac{1}{3}$ $\sqrt{7}$

9. 6.545 6.544 6.546 6.549

10. .285 29% 28% .287

Example 3

Arrange the values in **descending** order (**descending: greatest to least)**

-3.5 -3 0 -7 3.5

Negative values are always less than positive values
The "larger" the negative number, the less value it holds
Ex: -6 < -5

Answer

3.5 0 -3 -3.5 -7

Practice 4: Arrange the values in descending order

1. .006 .03% -14% .01

2. -16 -42 -8 -35

3. $\sqrt{14}$ 3.73 378% $\frac{11}{3}$

4. $\frac{1}{3}$ $\frac{1}{2}$ $\frac{1}{8}$ $\frac{1}{4}$

5. 2.060 2.059 2.606 2.509

6. -3 -.45 -2 -.453

7. 25% $\frac{1}{5}$ .30 .301

8. $\frac{2}{3}$ $\frac{4}{5}$ $\frac{1}{2}$ $\frac{2}{6}$

9. -.65 -.649 -.651 -.64993

10. .285 29% 28% .287

Review

1. Identify the largest value: -2 $-\frac{1}{2}$ -45% $-\frac{2}{3}$

2. Arrange the values in descending order: 65% $\frac{12}{25}$ 0.065 $\sqrt{.36}$

3. Identify the smallest value: $\frac{2}{3}$ $\frac{1}{6}$ $\frac{2}{9}$ $\frac{1}{8}$

4. Arrange the values in ascending order: 6.605 6.62 6.6 6.6052

5. Arrange the values in descending order: -4 -10 -2 -5

6. Identify the largest value: 23% 0.235 250% $\frac{29}{125}$

7. Identify the smallest value: 1.892 165% $\sqrt{2}$ $\sqrt{3}$

8. Arrange the values in ascending order: -2 $-\frac{1}{2}$ -45% $-\frac{2}{3}$

9. Identify the smallest value: 0.65% 0.65 $\frac{65}{1000}$ 65%

10. Arrange the values in descending order: $1\frac{2}{3}$ $\frac{4}{3}$ $\frac{7}{8}$ $1\frac{2}{5}$

11. Identify the largest value: 36% $\sqrt{.25}$ $\frac{2}{5}$ .364

12. Arrange the values in ascending order: 7.2% .082 12.5% .0823

13. Arrange the values in descending order: -23 -32 -30 -25

14. Identify the largest value: $\frac{1}{2}$ $\frac{2}{5}$ $\frac{3}{7}$ $\frac{4}{6}$

15. Identify the smallest value: 65% $\frac{12}{20}$ 0.065 $\sqrt{.36}$

16. Arrange the values in descending order: : 36% $\sqrt{.25}$ $\frac{2}{5}$ .364

17. Arrange the values in ascending order: $\frac{1}{2}$ $\frac{2}{5}$ $\frac{3}{7}$ $\frac{4}{6}$

18. Identify the smallest value: $\sqrt{5}$ 224% $\frac{53}{25}$ 228%

19. Identify the largest value: 7.2% .082 12.5% .0823

20. Arrange the values in ascending order: $\sqrt{5}$ 224% $\frac{53}{25}$ 228%

Answer Key

Practice 1

1. $\frac{9}{5}$

2. 68%

3. 2.689

4. $\frac{1}{2}$

5. $\sqrt{4}$

6. -0.4

7. -1.8660

8. $\sqrt{7}$

9. 6.549

10. 29%

Practice 2

1. -14%

2. -42

3. $\frac{11}{3}$

4. $\frac{1}{8}$

5. 2.059

6. -3

7. $\frac{1}{5}$

8. $\frac{2}{6}$

9. -0.651

10. $\sqrt{25}$

Practice 3

1. 1.728 $\sqrt{3}$ 1.734 $\frac{9}{5}$

2. $\sqrt{.25}$ .60 $\frac{2}{3}$ 68%

3. 2.654 2.664 2.688 2.689

4. -4 -3 -2 0 $\frac{1}{2}$

5. -$\sqrt{4}$ $\frac{3}{2}$ 152% $\sqrt{4}$

6. -5 -3 -$\frac{1}{2}$ -0.4

7. -1.869 -1.8668 -1.8667 -1.8660

8. $\frac{1}{3}$ 0.345 35% $\sqrt{7}$

9. 6.544 6.545 6.546 6.549

10. 28% .285 .287 29%

Practice 4

1. .01 .006 .03% -14%

2. -8 -16 -35 -42

3. 378% $\sqrt{14}$ 3.73 $\frac{11}{3}$

4. $\frac{1}{2}$ $\frac{1}{3}$ $\frac{1}{4}$ $\frac{1}{8}$

5. 2.606 2.509 2.060 2.059

Practice 4

6. -.45 -.453 -2 -3

7. .301 .30 25% $\frac{1}{5}$

8. $\frac{4}{5}$ $\frac{2}{3}$ $\frac{1}{2}$ $\frac{2}{6}$

9. -.649 -.64993 -.65 -.651

10. 29% .287 .285 28%

Review

1. -45%

2. 65% $\sqrt{.36}$ $\frac{12}{25}$ 0.065

3. $\frac{1}{8}$

4. 6.6, 6.605, 6.6052, 6.62

5. -2 -4 -5 -10

6. 250%

7. $\sqrt{2}$

8. -2 $-\frac{2}{3}$ $-\frac{1}{2}$ -45%

9. 0.65%

10. $1\frac{2}{3}$ $1\frac{2}{5}$ $\frac{4}{3}$ $\frac{7}{8}$

11. $\sqrt{.25}$

12.7.2% .082 .0823 12.5%

13. -23 -25 -30 -32

14. $\frac{4}{6}$

15. 0.065

16. $\sqrt{.25}$ $\frac{2}{5}$ .364 36%

17. $\frac{2}{5}$ $\frac{3}{7}$ $\frac{1}{2}$ $\frac{4}{6}$

18. $\frac{53}{25}$

19. 12.5%

20. $\frac{53}{25}$ $\sqrt{5}$ 224% 228%

Convert Within and Between Standard and Metric Systems

You will learn:

How to convert within metric and standard systems

How to convert between metric and standard systems

Study Tips

Read and study EVERY example problem.

Complete EVERY practice problem.

Check to make sure all answers are correct.

Go back to correct the questions you answered incorrectly.

If you don't receive at least an 80% on the review, go back and review the topic.

Convert Within Standard and Metric Systems

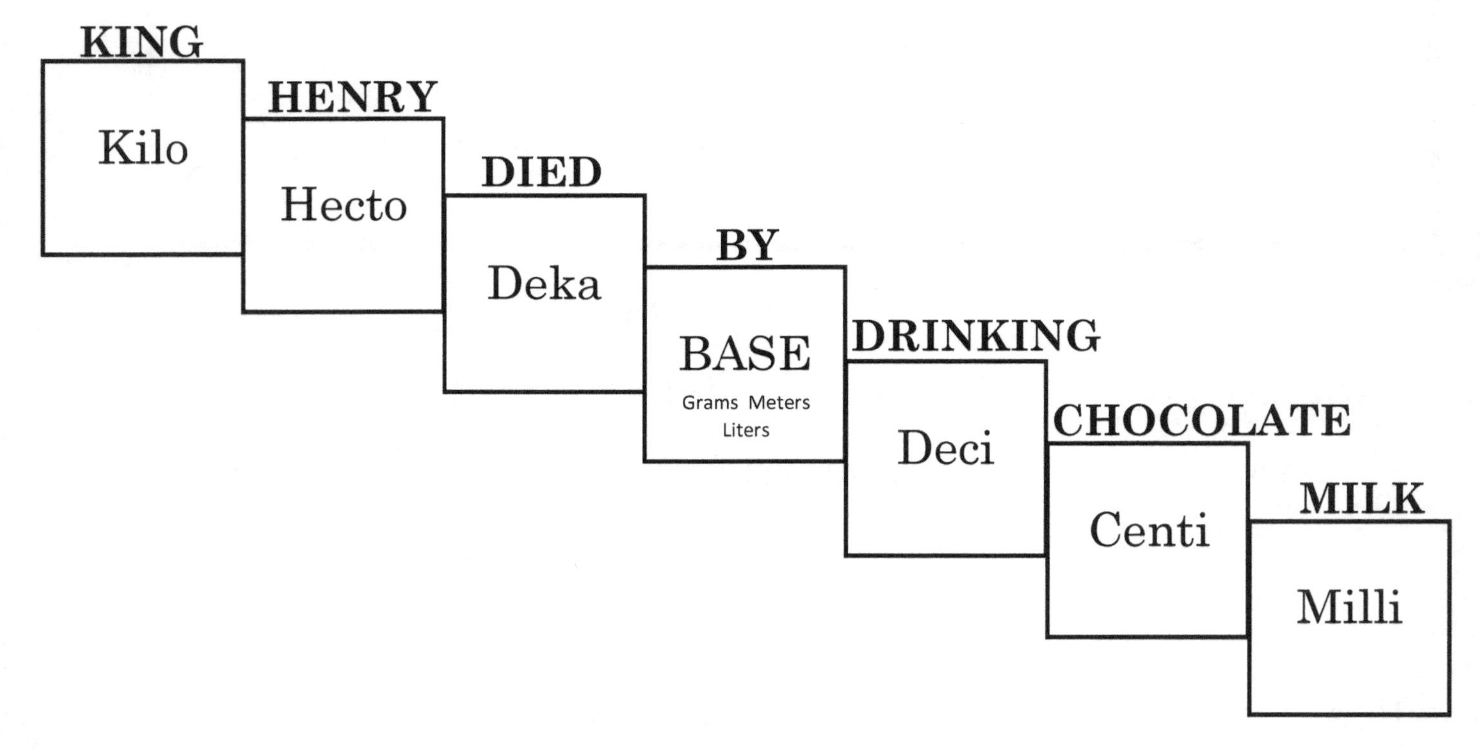

BASE UNITS

GRAMS **LITERS** **METERS**

Grams	Liters	Meters
kg: Kilogram	kl: Kiloliter	km: Kilometer
hg: Hectogram	hl: Hectoliter	hm: Hectometer
dkg: Dekagram	dkl: Dekaliter	dkm: Dekameter
g: Gram	l: Liter	m: Meter
dg: Decigram	dl: Deciliter	dm: Decimeter
cg: Centigram	cl: Centiliter	cm: Centimeter
mg: Milligram	ml: Milliliter	mm: Millimeter

KISS IT!

Step 1: Start at the decimal

Step 2: Move the decimal left or right

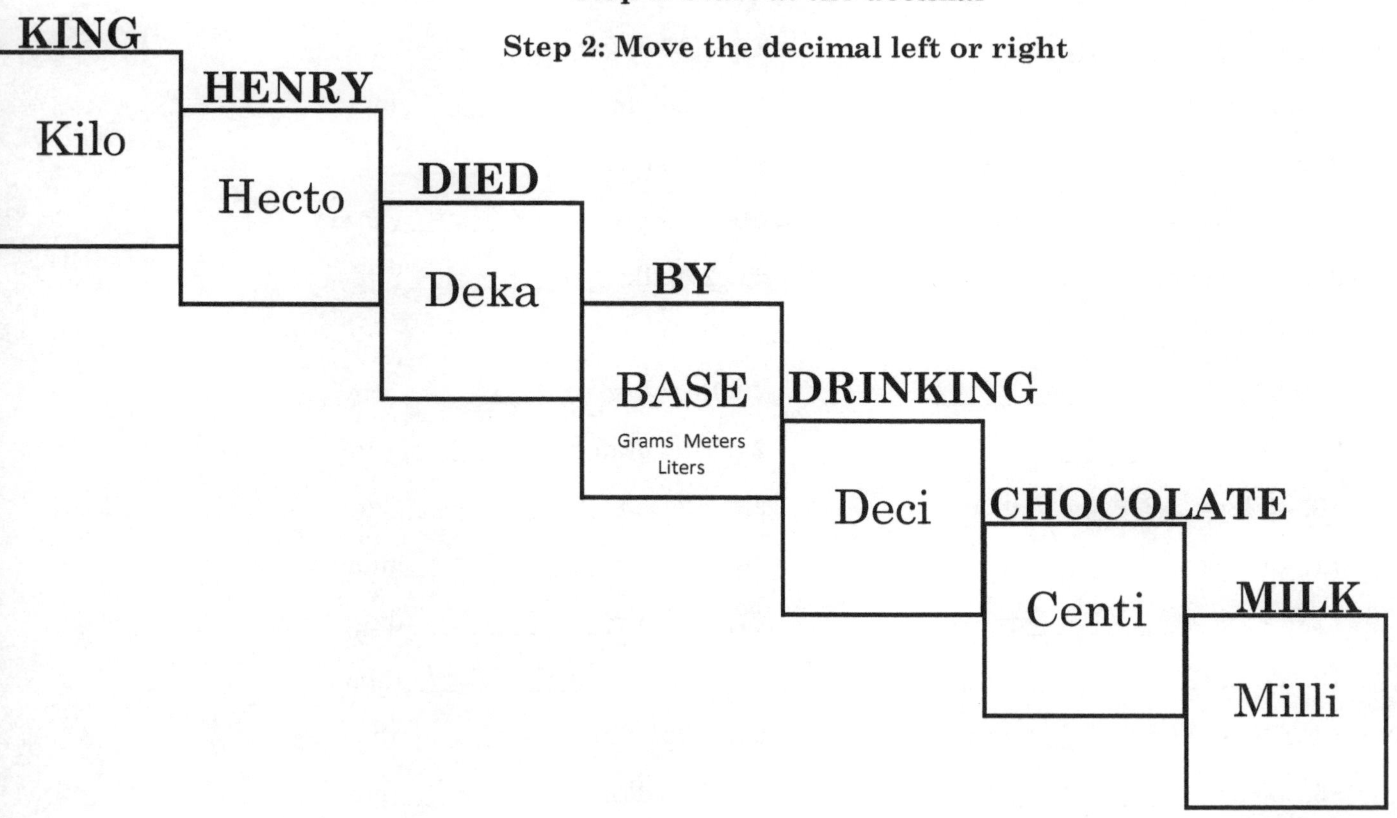

This method requires you to do just as the name states: move the decimal. This concept requires you to move the decimal left or right depending on the conversion you are required to complete. For example, if you are required to convert from kilograms to grams then you would count from kilo to the base (3 spaces). You would then move the decimal three spaces to the right

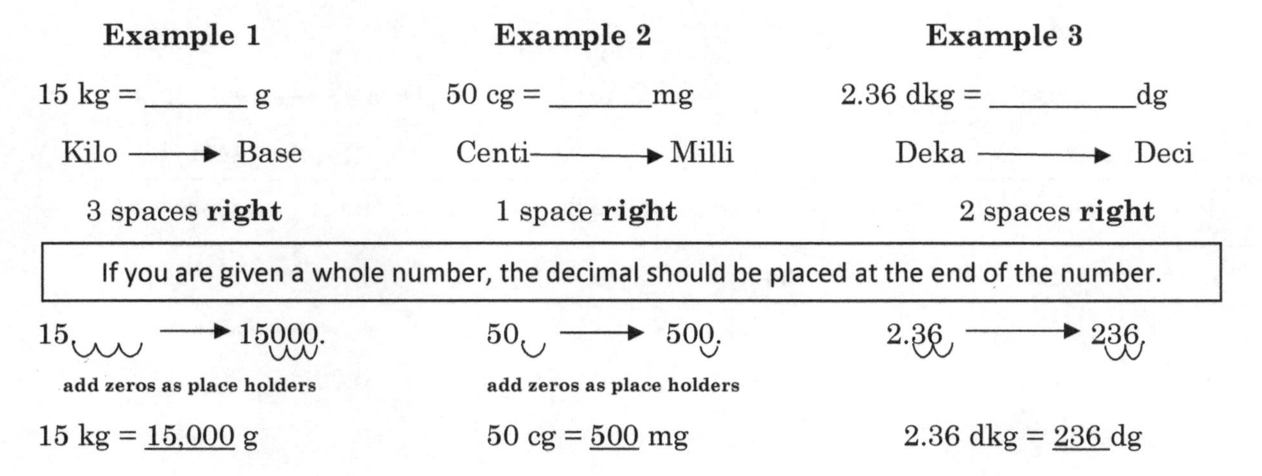

Practice 1

1. .638 kg = ________ g

2. 89.65 g = ________ cg

3. 3.5 cm = ________ mm

4. .00652 kg = ________ mg

5. .0032 kL = ________ ml

6. 50,000 m = ________ dm

7. 58 dkl = ________ cL

8. 0.62453 kL = ________ ml

9. 20 m = ________ cm

10. 45 kg = ________ cg

11. 45.3 cg = ________ mg

12. 5.62 g = ________ cg

13. .0032 cL = ________ ml

14. .7865 km = ________ mm

15. 35 g = ________ cg

16. .62 L = ________ ml

17. 3.28 dkg = ________ cg

18. .0287 m = ________ dm

19. 2,384 kg = ________ cg

20. 10 dkm = ________ mm

21. 273 dkl = ________ dl

22. 3 kg = ________ g

23. 24.7 km = ________ cm

24. .034 dkm = ________ m

25. 3.21 hm = ________ dm

26. 19 cm = ________ mm

27. .03 kg = ________ dkg

28. 2.2 m = ________ mm

29. .0025 km = ________ dm

30. .028 dkm = ________ mm

Example 4

125 ml = ______ l

Milli ⟶ Base

3 spaces **left**

Example 5

230 mg = ______ cg

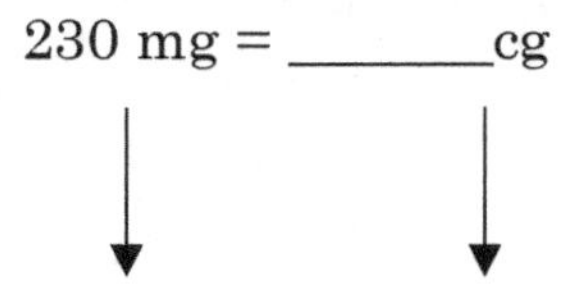

Milli ⟶ Centi

1 space **left**

Example 6

0.74 dg = ________ dkg

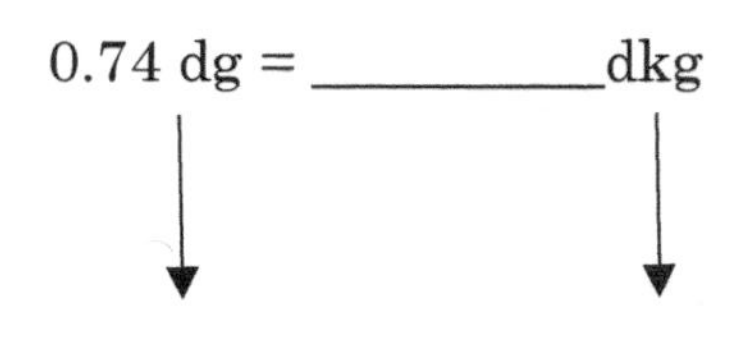

Deci ⟶ Deka

2 spaces **left**

If you are given a whole number, the decimal should be placed at the end of the number.

125. ⟶ .125

125 ml = .125 l

230. ⟶ 23.0

230 mg = 23 cg

0.74 ⟶ .0074

add zeros as place holders

0.74 dg = .0074 dkg

Practice 2

1. 625 g = ________ kg

2. 32.5 mm = ________ cm

3. 32 L = ________ kl

4. 898 ml = ________ kl

5. 56 cm = ________ m

6. 5640 cg = ________ kg

7. .6232 ml = ________ l

8. 5.62 cg = ________ g

9. .03 dm = ________dkm

10. 2.12 mm = ________cm

11. .32 dkl = ________kl

12. .029 dkm = ________ km

13. 2,945 mm = ________m

14. 273 dg = ________kg

15. 3 g = ________dkg

16. 89.65 mg = ________ g

17. .00652 g = ________dkg

18. 50,000 cm = ________km

19. .623 ml = ________dl

20. .0032 mg = ________g

21. 786.5 m = ________km

22. 45.3 mg = ________cg

23. 35 mg = ________dg

24. 12 dg = ________dkg

25. 180 l = ________kl

26. .923 mm = ________m

27. 2.73 m = ________km

28. .0038 ml = ________l

29. 7,543 g = ________dkg

30. 2 m = ________km

Practice 3

1. .0032 cL = ________ ml

2. 6232 ml = ________ l

3. 45.3 mg = ________cg

4. 19 cm = ________mm

5. .0032 l = ________ kl

6. 3.21 hm = ________dm

7. .028 dkm =________mm

8. 50,000 cm = ________km

9. .034 dkm =________m

10. 89.65 mg = ________ g

11. 2.12 mm = ________cm

12. 3.5 cm = ________ mm

13. 2,384 kg = ________ cg

14. .623 ml = ________dl

15. 5.62 cg = ________ g

16.10 dkm = ________ mm

17. 32.5 mm = ________ cm

18. .7865 mm = ________km

19. 2.2 m = ________mm

20. 24.7 km = ________cm

Answer Key

Practice 1

1. 638	**7.** 58,000	**13.** .032	**19.** 238,400,000	**25.** 3,210
2. 8,965	**8.** 624,530	**14.** 786,500	**20.** 100,000	**26.** 190
3. 35	**9.** 2,000	**15.** 3,500	**21.** 27,300	**27.** 3
4. 6,520	**10.** 4,500,000	**16.** 620	**22.** 3,000	**28.** 2,200
5. 3,200	**11.** 453	**17.** 3,280	**23.** 2,470,000	**29.** 25
6. 500,000	**12.** 562	**18.** .287	**24.** .34	**30.** 280

Practice 2

1. .625	**7.** .0006232	**13.** 2.945	**19.** .00623	**25.** .180
2. 3.25	**8.** .0562	**14.** .0273	**20.** .0000032	**26.** .000923
3. .032	**9.** .0003	**15.** .3	**21.** .7865	**27.** .00273
4. .000898	**10.** .212	**16.** .08965	**22.** 4.53	**28.** .0000038
5. .56	**11.** .0032	**17.** .000652	**23.** .35	**29.** 754.3
6. .05640	**12.** .00029	**18.** .5	**24.** .12	**30.** .002

Practice 3

1. .032	**6.** 3,210	**11.** .212	**16.** 100,000
2. 6.232	**7.** 280	**12.** 35	**17.** 3.25
3. 4.53	**8.** .5	**13.** 238,400,000	**18.** .0000007865
4. 190	**9.** .34	**14.** .00623	**19.** 2,200
5. .0000032	**10.** .08965	**15.** .0562	**20.** 2,470,000

CONVERT BETWEEN STANDARD AND METRIC UNITS

KISS IT!

Step 1: Set up Ratio

Step 2: Multiply or Divide

Let the set-up guide you!

Conversions come down to a matter of multiplying or dividing, and the key is understanding when to perform each operation. The **MISTAKE** that many people tend to make is guessing which operation to perform or assuming that one operation is more common than another. This method is a sure to give you correct answer **EVERY TIME!**

Example 1

25 yards = __________ meters

(1 yard = .915 meters)

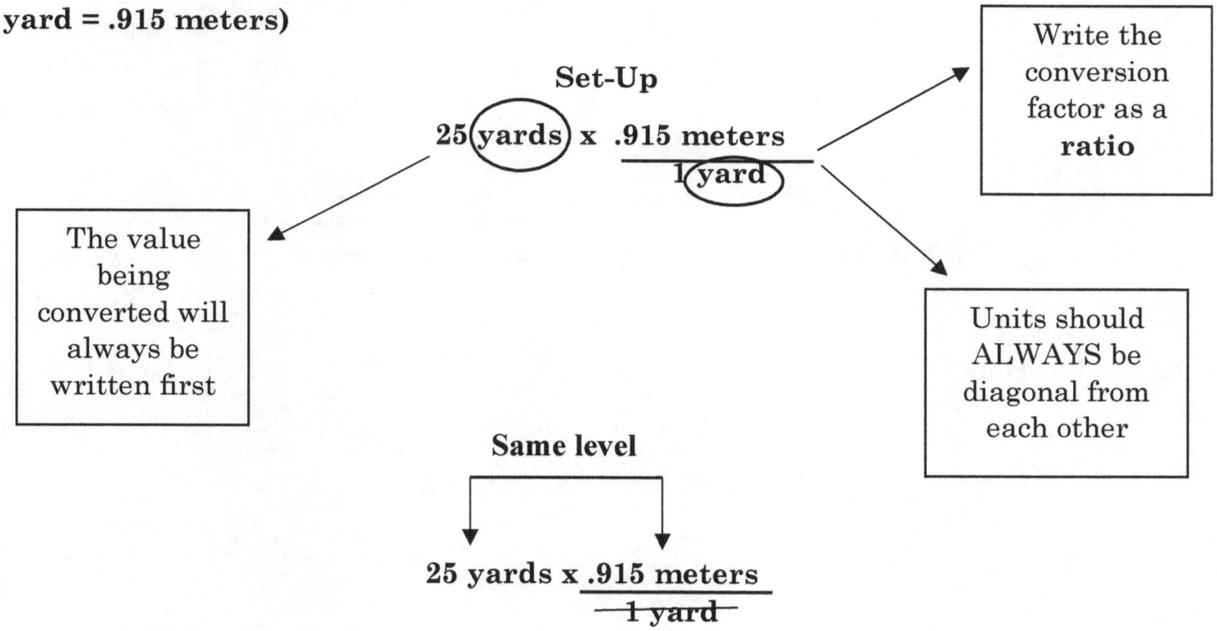

Set-Up

25 yards x .915 meters / 1 yard

Write the conversion factor as a **ratio**

The value being converted will always be written first

Units should ALWAYS be diagonal from each other

Same level

25 yards x .915 meters / ~~1 yard~~

* Same level means **MULTIPLY**

25 x .915 = 22.875

25 yards = 22.875 meters

Example 2

128 centimeters = __________inches

(1 inch = 2.54 centimeters)

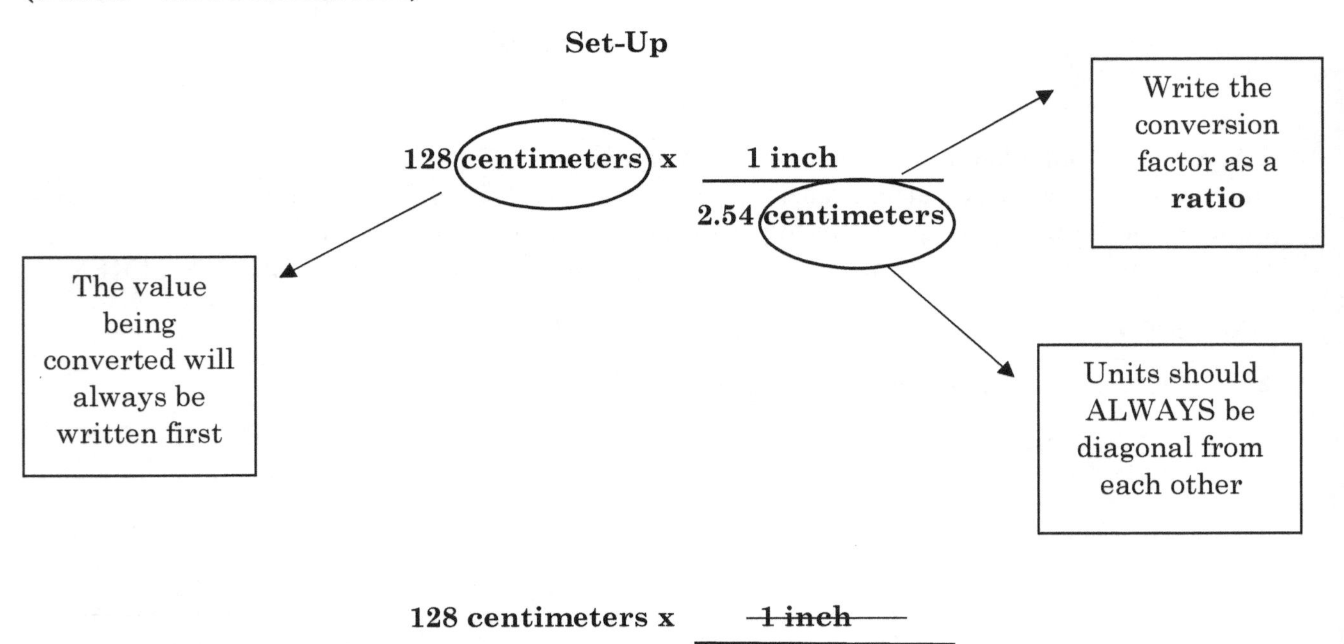

$$128 \text{ centimeters} \times \frac{\text{1 inch}}{2.54 \text{ centimeters}}$$

Different levels

*Different level means **DIVIDE**

128 ÷ 2.54 = 50.39

128 centimeters = 50.39 inches

Practice 4

1. 16 lbs. = __________ kg **(1 pound = 0.45 kilogram)**

2. 65 g = ____________ oz **(1 ounce = 28.35 grams)**

3. 23,760 feet = _________ miles **(1 mile = 5,280 feet)**

4. 62.8 miles = ___________km **(1 mile = 1.609 kilometers)**

5. 49.14 L = ____________ gallons **(1 gallon = 3.78 Liters)**

6. 14.4 kg = ____________ lbs. **(1 pound = 0.45 kilogram)**

7. 2,000 oz = ___________ tons **(1 Ton = 32,000 ounces)**

8. 52 meters = ___________yds **(1 yard = 0.914 meters)**

9. 15 feet = ____________meters **(1 foot = 0.305 meters)**

10. 4.8 yards = ____________ inches

11. 6.9 feet = ____________yards

12. 25°F = ______________ °C **(C = $\frac{5}{9}$ (F – 32))**

13. 125 oz = _____________ lbs. **(1 lb. = 16 oz)**

14. $1\frac{2}{5}$ miles =______________ feet **(1 mile = 5,280 feet)**

15. 26 mL = ________________ tbsp **(1 tbsp = 15 mL)**

16. 12.67 L = _______________ gallons **(1 gallon = 3.78 L)**

17. 12°C = ______________ °F **($F = \frac{9}{5}C + 32$)**

18. 32.8 cm = _____________inches **(1 inch = 2.54 cm)**

19. $4\frac{1}{4}$ ounces = _____________grams **(1 ounce = 28.35 grams)**

20. 6.34 miles = _______________km **(1 mile = 1.609 kilometers)**

21. 76°F = ______________ °C **((C = $\frac{5}{9}$ (F – 32))**

22. 589 yards = _____________miles **(1 mile = 1,760 yards)**

23. 0.85 tons = ______________ ounces **(1 Ton = 32,000 ounces)**

24. 75°C = ______________ °F **($F = \frac{9}{5}C + 32$)**

25. 8 fluid ounces = ______________mL **(1 fluid ounce = 29.57 mL)**

Review

1. 0.435 g = ____________mg

2. 2.45 mg = ____________kg

3. 0.02 dkg = ___________cg

4. 1.98 l = ______________ml

5. 945 kg = _____________g

6. 21.3 dl = _____________l

7. 24.7 km = _____________cm

8. 2.12 mm = _________cm

9. 7,543 g = ____________dkg

10. 19 cm = ______________mm

11. 378 cm = _____________inches **(1 inch = 2.54 cm)**

12. 23°C = ______________ °F **($F = \frac{9}{5}C + 32$)**

13. 0.46 miles = _______________km **(1 mile = 1.609 kilometers)**

14. 3.25 yards = ____________ inches

15. 98°F = _____________ °C **($C = \frac{5}{9}(F - 32)$)**

16. $7\frac{2}{3}$ ounces = _____________grams **(1 ounce = 28.35 grams)**

17. 2.34 feet = ____________yards

18. 28 kg = ____________ lbs. **(1 pound = 0.45 kilogram)**

19. 39 lbs. = __________ kg **(1 pound = 0.45 kilogram)**

20. 5.9 mL = ________________ tbsp **(1 tbsp = 15 mL)**

21. 5990 cg = __________ kg

22. 0.75 feet = ____________inches

23. 298 dkl = _____________ dl

24. 1.53 feet = ____________meters **(1 foot = 0.305 meters)**

25. 2.97 g = _____________ kg

26. 43°C = ______________ °F **($F = \frac{9}{5}C + 32$)**

27. 0.273 dm = _____________ mm

Practice 4

1. 7.2 kg
2. 2.29 oz
3. 4.5 miles
4. 101.05 km
5. 13 gallons
6. 32 lbs.
7. 0.0625 tons
8. 56.89 yds
9. 4.575 m
10. 172.8 in
11. 2.3 yds
12. -3.9°C
13. 7.81 lbs.
14. 7,392 ft
15. 1.73 tbsp
16. 3.35 gallons
17. 53.6°F
18. 12.91 in
19. 120.49 g
20. 10.20 km
21. 24.4°C
22. .33 miles
23. 27,200 oz
24. 167°F
25. 236.56 mL

Review

1. 435
2. .00000245
3. 20
4. 1,980
5. 945,000
6. 2.13
7. 2,470,000
8. .212
9. 754.3
10. 190
11. 148.8 in
12. 73.4°F
13. .74 km
14. 117 in
15. 36.7°C
16. 217.35 g
17. .78 yards
18. 62.2 lbs.
19. 17.55 kg
20. .39 tbsp
21. 0.05990
22. 9 inches
23. 29,800
24. 0.4667
25. 0.00297
26. 109.4°F
27. 27.3

BLANK PAGE

Cluster 1 Review

You should know:

☐ How to round whole numbers and decimals

☐ How to simplify a radical

☐ How to choose an appropriate tool of measure

☐ How to estimate the mass, length, and width of everyday objects

☐ How to convert fractions to decimals

☐ How to convert fractions to percentages

☐ How to convert decimals to percentages

☐ How to convert decimals to fractions

☐ How to convert percentages to fractions

☐ How to convert percentages to decimals

☐ How to compare rational values

☐ How to identify the largest value in a data set

☐ How to identify the smallest value in a data set

☐ How to arrange values in ascending order

☐ How to arrange values in descending order

☐ How to convert within standard and metric systems

☐ How to convert between standard and metric systems

CLUSTER 1 REVIEW

1. Convert $\frac{6}{29}$ to a decimal

a. 4.84 b. .206 c. 4.83 d. .207

2. Identify the largest value

a. 8.684 b. 8.648 c. 4.83 d. .207

3. 6.6 mg = ______________ g

a. 6.6 b. 660 c. .0066 d. 6600

4. Estimate $\sqrt{8}$

a. 2.83 b. 2.82 c. 2.5 d. 2.9

5. Convert 125% to a decimal

a. 125 b. 1.25 c. 12.5 d. 1250

6. 348 oz = ______________lb. **(1lb = 16 oz)**

a. $21\frac{1}{4}$ b. 5568 c. $21\frac{3}{4}$ d. .046

7. Arrange the values in descending order: 42% $\sqrt{.36}$ $\frac{3}{8}$ .376

a. $\sqrt{.36}$ 42% .376 $\frac{3}{8}$

b. 42% .376 $\sqrt{.36}$ $\frac{3}{8}$

c. $\sqrt{.36}$ 42% $\frac{3}{8}$ .376

d. $\frac{3}{8}$ .376 42% $\sqrt{.36}$

8. 1.68 kg = ____________ cg

a. 168 b. 16.8 c. 16,800 d. 168,000

9. 90.72 L = ____________gallons **(1 gallon = 3.78 L)**

a. 343 b. 24 c. .04 d. 25

10. Which unit is best when estimating the weight of a small child?

a. Grams b. liters c. feet d. kilograms

11. Convert $\frac{3}{8}$ to a percentage

a. .375% b. 2.67% c. 37.5% d. 267%

12. Identify the smallest value: -12 -13 -14 -10

a. -12 b. -13 c. -14 d. -10

13. 206 L = ___________ kL

a. .206 b. 20.6 c. 2,060 d. 20,600

14. $2\frac{1}{4}$ yards = ____________ inches

a. 81 b. 6.75 c. 18 d. 20.25

15. Which unit is best when estimating the length of a coffee table?

a. Feet b. inches c. ounces d. liters

16. Arrange the values in ascending order: $-\frac{1}{4}$ $\sqrt{9}$ -2 160%

a. $\sqrt{9}$ 160% $-\frac{1}{4}$ -2
b. -2 $-\frac{1}{4}$ 160% $\sqrt{9}$
c. $\sqrt{9}$ 160% -2 $-\frac{1}{4}$
d. $-\frac{1}{4}$ -2 160% $\sqrt{9}$

17. .086 m = _______________ mm

a. 8.6 b. .86 c. 860 d. 86

18. Convert 0.85 to a fraction in simplest form

a. $\frac{17}{20}$ b. $\frac{85}{1000}$ c. $\frac{85}{10}$ d. $\frac{20}{17}$

19. 6.89 miles = ____________ km **(1 mile = 1.609 km)**

a. 4.28 b. 11.09 c. 4.29 d. 11.8

20. Estimate the product 466 × 52

a. 25,000 b. 20,000 c. 23,000 d. 24,000

Answer Key

1. D

2. A

3. C

4. A

5. B

6. C

7. A

8. D

9. B

10. D

11. C

12. C

13. A

14. A

15. A

16. B

17. D

18. A

19. B

20. A

Cluster 1 Review Break Down

Apply estimation strategies and Rounding Rules to Real-World Problems

Question 4 ☐

Question 10 ☐

Question 15 ☐

Question 20 ☐

Convert Among Non-Negative Fractions, Decimals, and Percentages

Question 1 ☐

Question 5 ☐

Question 11 ☐

Question 18 ☐

Compare and Order Rational Numbers

Question 2 ☐

Question 7 ☐

Question 12 ☐

Question 16 ☐

Convert Within and Between Standard and Metric Systems

Question 3 ☐

Question 6 ☐

Question 8 ☐

Question 9 ☐

Question 13 ☐

Question 14 ☐

Question 17 ☐

Question 19 ☐

BLANK PAGE

Cluster 2

Solve Real-World Problems Involving Ratios and Rate of Change

How to write the ratio between two quantities

How to identify equivalent ratios

How to solve for missing values given a ratio

Solve Real-World Problems Involving Proportions

How to solve proportions with whole numbers

How to solve proportions with fractions and decimals

How to solve proportions containing values with different units

Solve Real-World Problems Involving Percentages

How to calculate percentage increase

How to calculate percentage decrease

How to calculate tax, tip & discount

How to calculate percent markups

How to calculate percent markdowns

How to solve percent proportions

BLANK PAGE

Solve Real-World Problems Involving Ratios and Rate of Change

You will learn:

How to write the ratio between two quantities

How to identify equivalent ratios

How to solve for missing values given a ratio

Study Tips

Read and study EVERY example problem.

Complete EVERY practice problem.

Check to make sure all answers are correct.

Go back to correct the questions you answered incorrectly.

If you don't receive at least an 80% on the review, go back and review the topic.

Ratios

Ratios are used to show the relationship between two quantities

Ratios can be written three ways:

using the word "to"	example: 2 to 3
using a colon (:)	example: 2:3
as a fraction	example: $\frac{2}{3}$

IDENTIFYING RATIOS FROM GIVEN INFORMATION

KISS IT

Step 1: Identify the quantities being compared

Step 2: Write your ratio

Example 1

There are 28 marbles in a jar: 7 green, 11 blue, 5 red and 5 orange. What is the ratio of blue marbles to red marbles?

Step 1: Identify the quantities being compared

→ blue marbles **to** red marbles

Step 2: Write your ratio

11 to 5 or $\frac{11}{5}$ or 11:5

Example 2

A candy machine contains 20 jolly ranchers, 23 blow pops, and 42 gumballs. What is the ratio of blow pops and jolly ranchers to the total number of pieces in the machine?

Step 1: Identify the quantities being compared

→ Blow pops & jolly ranchers **to** total pieces

Blow pops + Jolly Ranchers **to** Total (blow pops + jolly ranchers + gumballs)

23 + 20 = 43

23 + 20 + 42 = 85

Step 2: Write your ratio

43 to 85 or $\frac{43}{85}$ or 43:85

The word "to" acts as a TRIGGER WORD: whatever comes before "to" is written first in the ratio & whatever comes after "to" is written second

Practice 1

1. A bag contains 10 blue marbles, 7 red marbles, and 8 purple marbles
a. What is the ratio of red marbles to blue marbles?
b. What is the ratio of purple marbles to the total number of marbles?

2. In a class of 20 students, there are 4 basketball players, 7 football players, 1 soccer player, and the remaining students are the on the track team.
a. What is the ratio of football players to students who run track?
b. What is the ratio of basketball and soccer players to the total number of students who run track?

3. Aaliyah found 25 quarters, 18 dimes, 6 nickels, and one penny in her purse.
a. What is the ratio of pennies to the total number of coins?
b. What is the ratio of quarters to dimes and nickels?

4. Robert has 3 watches, 2 bracelets, 2 pairs of sneakers, 3 pairs of casual shoes, and 1 ring in his suitcase.
a. What is the ratio of shoes to jewelry?
b. What is the ratio of watches and sneakers to the total number of items in his bag?

5. The local zoo houses 43 monkeys, 6 lions, 4 tigers, 6 bears, and 80 birds.
a. What is the ratio of lions and tigers to monkeys?
b. What is the ratio of birds to the total number of animals in the zoo?

6. A jewelry box contains 16 earrings, 8 bracelets, 7 necklaces, and 2 watches.
a. What is the ratio of bracelets to watches?
b. What is the ratio of bracelets and watches to necklaces and earrings?

7. Lola's book collection includes 12 non-fiction books, 15 fiction books, and 5 comic books
a. What is the ratio of comic books to fiction books?
b. What is the ratio of non-fiction books to fiction books?

8. The local college's graduating class includes 29 nursing majors, 82 communications majors, 59 psychology majors, 49 criminal justice majors, and 83 education majors.
a. What is the ratio of criminal justice majors to the total number of graduates?
b. What is the ratio of nursing and communications majors to education majors?

9. A bag contains 75 marbles: 20 red, 32 blue, and the remaining are green.
a. What is the ratio of green marbles to red marbles?
b. What is the ratio of red marbles to the total number of marbles in the bag?

10. A bag contains 63 pieces of candy: 21 red, 32 blue, and 10 green.
a. What is the ratio of red pieces to the total number of pieces in the bag?
b. What is the ratio of blue and green pieces to red pieces in the bag?

IDENTIFYING EQUIVALENT RATIOS

KISS IT

Step 1: Cross Multiply

Example 3

The ratio of boys to girls at Campbell Middle School is 25:100. Holly Hill Middle School has an equivalent ratio of boys to girls. Which option represents the ratio of boys to girls at Holly Hill Middle?

a. $\frac{1}{5}$ b. $\frac{5}{100}$ c. $\frac{5}{10}$ d. $\frac{5}{20}$

Step 1: Cross multiply

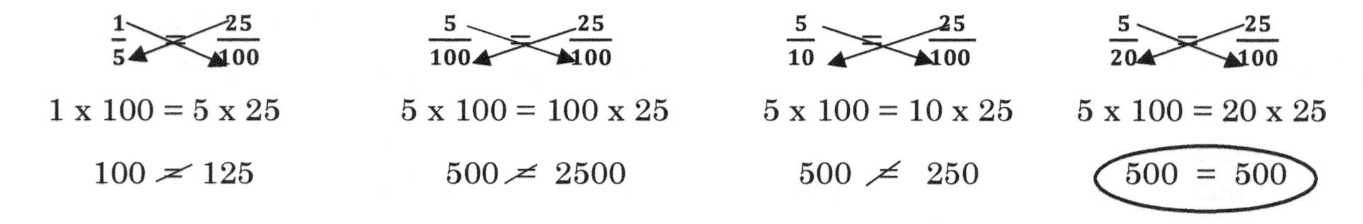

a. $\frac{1}{5}$ b. $\frac{5}{100}$ c. $\frac{5}{10}$ d. $\frac{5}{20}$

Answer: D

Example 4

The ratio of base to height on a triangle is 3 to 4. Identify an equivalent ratio from the options given below

a. $\frac{9}{16}$ b. $\frac{1}{4}$ c. $\frac{15}{20}$ d. $\frac{3}{5}$

Step 1: Cross multiply

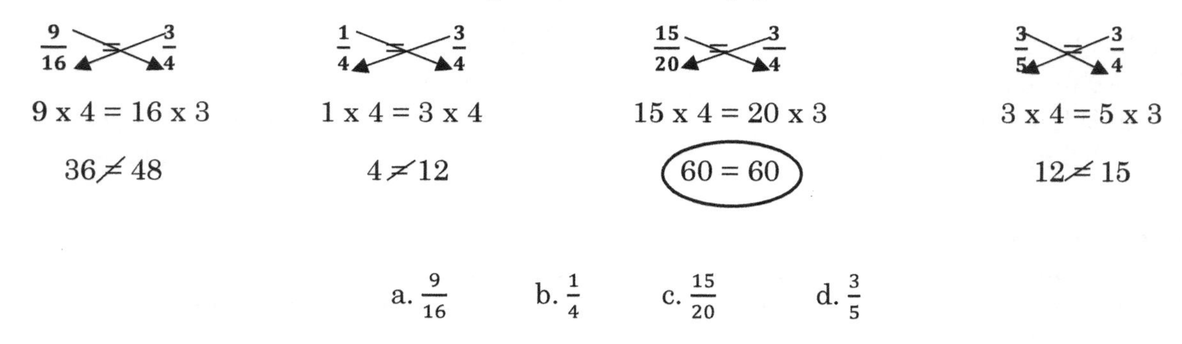

Answer: C

Practice 2

1. Identify the ratio that is equivalent to 2:5

a. 5:2 b. 4:7 c. 10:25 d. 4:1

2. Is the ratio 1:3 equivalent to the ratio 6:8

a. yes **b.** no

3. The Dolphin's ratio of wins to losses is 5 to 2. The Cowboys have an equivalent ratio of wins to losses, which option represents this ratio?

a. $\frac{2}{5}$ **b.** $\frac{2.5}{2}$ **c.** $\frac{10}{4}$ **d.** $\frac{8}{5}$

4. The ratio of dogs to cats at the Nearburg Animal Shelter is 5:4. Every shelter in the city has an equivalent ratio of dogs to cats, except for the Leesburg Animal Shelter. Identify the ratio that represents dogs and cats at the Leesburg shelter.

a. 15 to 12 **b.** 50 to 40 **c.** 7 to 6 **d.** 25 to 20

5. The ratio of base to height on a parallelogram is $\frac{1}{4}$. Identify an equivalent ratio.

a. $\frac{4}{1}$ **b.** $\frac{10}{40}$ **c.** $\frac{4}{4}$ **d.** $\frac{20}{40}$

6. Identify the ratio that is equivalent to 10:3

a. $\frac{3}{10}$ b. $\frac{20}{13}$ c. $\frac{20}{6}$ d. $\frac{33}{1}$

7. The ratio of men to women at the local gym is 5 to 1. A gym nearby has an equivalent ratio of men to women as well. If there is a total of 250 men at this gym, how many women are there?

a. 50 b. 250 c. 246 d. 60

8. Are the ratios 2 to 5 and 5 to 2 equivalent

a. yes b. no

9. The ratio of white socks to black socks in Eric's top drawer is 4:7. If his brother, Marcus, has an equivalent ratio of white socks to blacks, identify a possible ratio

a. $\frac{7}{4}$ b. $\frac{20}{35}$ c. $\frac{8}{11}$ d. $\frac{1}{7}$

10. The ratio of lace frontals to closures in stock at Diamond Lengths is 1:4. If an equivalent ratio of frontals to closures includes 20 frontals, how many closures would be included?

a. 4 b. 20 c. 60 d. 80

IDENTIFYING MISSING VALUES GIVEN A RATIO

KISS IT

***Trigger: 3 values & 2 different units**

Step 1: Set up units

Step 2: Insert values

Step 3: Cross multiply, then divide

Example 5

The ratio of cats to dogs is 16:4. There are 96 cats at the pet store, how many dogs are there?

*Note: This set up is identical to proportions

Step 1:Set up units $\frac{cats}{dogs} = \frac{cats}{dogs}$

Step 2: Insert values $\frac{\textbf{16}\text{ cats}}{\textbf{4}\text{ dogs}} = \frac{\textbf{96}\text{ cats}}{\text{dogs}}$

Step 3: Solve $\frac{\textbf{16}\text{ cats}}{\textbf{4}\text{ dogs}}$ *multiply* $\frac{\textbf{96}\ cats}{\text{dogs}}$

4 x 96 = 384

After you multiply across, divide by the number that is left over

384 ÷ **16** = 24

Practice 3

1. The ratio of males to females in 8th grade is 2:3. If there are 50 males in the 8th grade, how many females are there?

2. Khloe has 20 orange marbles in a jar. If the ratio of blue marbles to orange marbles is 2:5, how many blue marbles are in the jar?

3. The ratio of students to teachers at Bethune-Cookman University is 13:1. If there are 39 students in a classroom, how many teachers should be present?

4. The ratio of Democrats to Republicans in Luke's office is 12 to 9. If 81 of his coworkers label themselves as Republicans, how many are Democrats?

5. The ratio of chickens to pigs on a farm is $\frac{4}{9}$. If there are 10 chickens total, how many pigs are there?

6. Ralph caught 30 red snappers today during his weekly fishing trip. If the ratio of red snappers to sheepshead is 2:3, how many sheepsheads did he catch?

7. Devin received 250 custom shoe orders this past Christmas. If the ratio of custom shoe orders to custom shirt orders is 10:5, how many custom shirt orders did he receive?

8. The ratio of male to female models on the runway is 1:5. If there are 30 female models, how many males are there?

9. The ratio of apples to oranges at the breakfast bar is 3:1. If there are 12 oranges at the bar, how many apples are there?

10. If it takes 10 people to pull a 20-ton truck, how many people will it take to pull 35 tons?

Review

1. The ratio of nurses to doctors at the local hospital is 10 to 1. If there are 40 doctors working in the hospital, how many nurses are there?

2. The ratio of nurses to doctors at the local hospital is 10:1. If the neighboring hospital has an equivalent number of nurses to doctors, identify a possible ratio that represents the scenario.

a. 1:10 b. 15:6 c. 80:8 d. 9:0

3. A bag contains 8 chocolate bars, 10 licorice sticks, 6 cupcakes, and 22 peppermints.

a. What is the ratio of chocolate bars to cupcakes and peppermints?

b. What is the ratio of peppermints to the total number of items in the bag?

4. The ratio of a triangle's base to height is 2:3. If the height of the triangle is 23, what is the length of its base?

5. There are 5 basketball players in Mr. Anderson's fifth period class. If the ratio of football players to basketball players in his class is 4:1, how many football players are there?

6. Identify the ratio that is not equivalent to 4:7

a. 8: 14 b. 7:4 c. 40:70 d. 20:35

7. The ratio of men to women who purchased new gym memberships for Christmas is 3:5. If 25 women purchased memberships, how many men purchased?

8. The traveling circus includes 23 clowns, 14 gymnasts, 82 animals, and 2 ring masters.

a. What is the ratio of animals to the total number of circus participants?

b. What is the ratio of gymnasts to animals?

9. For every teenager who does not own a cell phone, there are 100 who do. If 1,000 teenagers own a cell phone, how many do not?

10. Identify the ratio that is equivalent to 12:27

a. 27 to 12 b. 17 to 32 c. 7 to 22 d. 4 to 9

11. A garden contains 12 roses, 16 daisies, 4 tulips, and 10 sunflowers.

a. What is the ratio of roses to daisies?

b. What is the ratio of sunflowers to the total number of the flowers in the garden?

12. The ratio of male cheerleaders to female cheerleaders at the local high school is 1 to 5. If there are 4 males, how many cheerleaders are there in all?

13. The ratio of a trapezoid's base to height is 4 to 7. Identify a triangle with an equivalent ratio.

a. 12:21 b. 7:4 c. 16:21 d. 14:17

14. Ms. Baker's sixth period science class includes 15 boys and 17 girls.

a. What is the ratio of boys to girls.

b. What is the ratio of girls to the total number of students in the class?

15. Identify the ratio that is equivalent $\frac{2}{3}$

a. 3:2 b. 14:21 c. 7:8 d. 3:4

16. The ratio of the soccer team's wins to losses is 5 to 7. If they won 25 games, how many games did they play in all?

17. The ratio of men to women who drive sports cars is 3:1. If 33 women are recorded as driving sports cars, how many men are there?

18. Identify the ratio that is equivalent to 10:100

a. 2:20 b. 100:10 c. 20:120 d. 50:200

19. Stephanie's snack drawer contains 6 cookies, 5 chocolate bars, and 6 packs of pecans.

a. What is the ratio of cookies to pecans?

b. What is the ratio of chocolate bars to the total number of snacks in the drawer?

20. Brittany's dance class includes 5 sophomores, 7 juniors, and 8 seniors.

a. What is the ratio of sophomores to juniors and seniors?

b. What is the ratio of seniors to the total number of dancers?

Answer Key

Practice 1

1. a. 7:10

b. 8:25

2. a. 7:8

b. 5:8

3. a. 1:50

b. 25:24

4. a. 5:6

b. 5:11

5. a.10:43

b. 80:139

6. a. 4:1

b. 10:23

7. a. 1:3

b. 4:5

8. a. 49:302

b. 111:83

9. a. 23:20

b. 4:15

10. a. 1:3

b. 2:1

Practice 2

1. C

2. B

3. C

4. C

5. B

6. C

7. A

8. B

9. B

10. D

Practice 3

1. 75

2. 8

3. 3

4. 108

5. 22

6. 45

7. 125

8. 6

9. 36

10. 17.5

Review

1. 400

2. C

3. a. 2:7

b. 11: 23

4. 15.3

5. 20

6. B

7. 15

8. a. 82:121

b. 7:41

9. 10

10. D

11. a. 3:4

b. 5:21

12. 24

13. A

14. a.15:17

b. 17:32

15. B

16. 60

17. 99

18. A

19. a. 1:1

b. 5:17

20. a. 1:3

b. 2:5

BLANK PAGE

Solve Real-World Problems Involving Proportions

You will learn:

How to solve proportions with whole numbers

How to solve proportions with fractions and decimals

How to solve proportions containing values with different units

Study Tips

Read and study EVERY example problem.

Complete EVERY practice problem.

Check to make sure all answers are correct.

Go back to correct the questions you answered incorrectly.

If you don't receive at least an 80% on the review, go back and practice the topic.

KISS IT!

***Trigger: 3 values & 2 different units**

Step 1: Set up units

Step 2: Insert values

Step 3: Cross multiply, then divide

Example 1

Roman drives 150 miles on 15 gallons of gas. If he uses 25 gallons of gas, how far has he traveled.

Step 1 (set up units) $\frac{miles}{gallons} = \frac{miles}{gallons}$

Step 2 (insert values) $\frac{\mathbf{150}\ miles}{\mathbf{15}\ gallons} = \frac{miles}{\mathbf{25}\ gallons}$

Step 3 (solve) $\frac{\mathbf{150}\ miles}{\mathbf{15}\ gallons}$ ~~multiply~~ $\frac{miles}{\mathbf{25}\ gallons}$

150 x 25 = 3,750

Divide by the number that's left over

3,750 ÷ 15 = 250 miles

Practice 1

1. Lisa can drive 300 miles on 20 gallons of gas. If her tank can hold 35 gallons, how far can she travel on a full tank?

2. Benny can type 40 words per minute. If he types for 15 minutes, how many words will he have typed?

3. A dog can eat 3 bones in 2 hours. At this rate, how long will it take him to eat 10 bones?

4. For every 3 girls in the choir there are 2 boys. If there are 22 boys total, how many girls are there?

5. Every 5 plants require 1 gallon of water to stay hydrated. If Gloria used 4 gallons of water, how many plants can she hydrate?

6. The length of a rectangle is 9 inches, and its width is 6 inches. If a similar rectangle has a length of 30, what is its width?

7. The market's most efficient car is advertised as getting up to 150 mpg. If this is true, how far can this car travel on 7 gallons of gas?

8. Carter High School's cheerleaders can execute 10 stunts per quarter at a football game. At this rate, how many stunts will they have completed after 3 quarters?

9. For every acre of land the Jefferson's own, they have 20 horses and 30 cattle. If they own 8 acres of land, how many cattle do they own?

10. Alexis can type 50 words per minute. At this rate, how many words can she type in 30 minutes?

Example 2

Stephanie uses $3\frac{1}{2}$ cups of sugar to bake 5 red velvet cakes. How many cakes can she bake with 9 cups of sugar?

Step 1 (set up units) $\frac{cups}{cakes} = \frac{cups}{cakes}$

Step 2 (insert values) $\frac{3\frac{1}{2}\ cups}{5\ cakes} = \frac{9\ cups}{cakes}$

Step 3 (solve) $\frac{3.5\ cups}{5\ cakes}$ multiply $\frac{9\ cups}{cakes}$

*Note: If given a fraction, convert to a decimal

5 x 9 = 45

Divide by the number that's left over

45 ÷ 3.5 = 12.8 cakes

Practice 2

1. Annie uses $1\frac{3}{4}$ cups of flour to bake 2 lemon cakes. If she bakes 11 cakes for the annual Christmas party, how many cups of flour will she need?

2. Derrick can type $3\frac{2}{8}$ pages in $2\frac{1}{2}$ hours. At this rate, how many pages can he type in 4 hours?

3. The original dimensions of Alanah's favorite photo was 7 inches long by $5\frac{1}{2}$ inches wide. If the length of the enlarged picture is 10 inches, what is the width?

4. Tara purchased two lollipops for 85 cents. How much money will she spend if she purchases 25 lollipops?

5. Kendrick purchased one bag of grapes for $5.39. If she has $25, how many bags of grapes can she purchase?

6. A right triangle has a base of $8\frac{1}{2}$ inches and a height of $12\frac{1}{2}$ inches. If the height is reduced to 4 inches, how many inches is the new base?

7. You can buy $6\frac{7}{8}$ cartons of juice for $7.50. If Sheila spends $52.50, how many cartons of juice did she purchase?

8. For every 4 gallons of water she drinks, Ashley loses $\frac{3}{8}$ lb. If she drinks 14 gallons of water, how many pounds will she lose.

9. A rectangle has a length of $2\frac{5}{8}$ inches and a width of $1\frac{1}{4}$ inches. If the rectangle is enlarged, and its new length is 7 inches, what is its new width?

10. Alice can bake 4 pies in $2\frac{3}{4}$ hours. At this rate, how long will it take her to bake 9 pies?

Example 3

Angela can type 400 words in 8 minutes. At this rate, how many words can she type in 1 hour?

***Note: There are two different time units**

When this occurs, you will need to convert them both to the same unit. In this instance you can either convert minutes to hours or hours to minutes.

Step 1 (set up units) $\frac{words}{minutes} = \frac{words}{minutes}$

Step 2 (insert values) $\frac{400\ words}{8\ minutes} = \frac{words}{60\ minutes}$

***Note: 1 hour was converted to 60 minutes**

Step 3 (solve) $\frac{400\ words}{8\ minutes}$ ~~multiply~~ $\frac{words}{60\ minutes}$

400 x 60 = 24,000

Divide by the number that's left over

24,000 ÷ 8 = 3000 words

Practice 3

1. Angela can type 250 words in 12 minutes. At this rate, how many words can she type in 2 hours?

2. Courtney purchased 4 pairs of shoes this week. At this rate, how many pairs will she have purchased in 3 months?

3. It takes David 3 days to drink 2 gallons of water. At this rate, how many weeks will it take him to drink 15 gallons of water?

4. If students can master one standard every 4 days. How many weeks will it take to master 5 standards (not including weekends)?

5. Alexis can type 60 words per minute. At this rate, how many words can she type in three quarters of an hour?

6. A rectangle has a length of 6 inches and a width of 9 inches. If a similar rectangle has a length of 1 foot, how wide is it?

7. If 2 boxes of trash bags can last one month, how many individual trash bags can last for five months? (1 box contains 32 trash bags)

8. Reginald can type 45 words per minute. At this rate how many words can he type in one day?

9. Each student in Ms. Blair's math class uses two boxes of pencils each semester. How many individual pencils will her students use in one school year? (1 box = 12 pencils)

10. It is believed that it takes 21 days to form a habit. If this is true, how many months will it take a person to form 6 habits?

Review

1. Linda ran 6 miles in 2 hours. If she runs at the same rate, how long will it take her to run 24 miles?

2. If Jakara can bake forty-five cookies in thirty minutes, how many cookies can she bake in 90 minutes?

3. When Kevin travels home to Daytona Beach for spring break, it takes him $4\frac{1}{2}$ hours to drive 350 miles. Jacksonville is an additional 175 miles away, if he drives at the same rate how long will take him to arrive?

4. Barbara purchased 16 pairs of shoes in 4 weeks. If she continues to buy shoes at this rate, how many pairs will be she have purchased in 3 months?

5. Robert can type $3\frac{1}{4}$ pages in $2\frac{1}{2}$ hours. How many pages can he type in eleven hours?

6. Allen can draft 5 plays in 3 hours. How many hours will it take him to draft 32 plays?

7. Shank can paint 3 pairs of shoes per day. How many pairs can she paint in three weeks?

8. Raekia completed 4 photoshoots in $8\frac{1}{2}$ hours. At this rate, how many photoshoots can she complete in 19 hours?

9. A florist uses 3 pounds of roses in her specialty vase. How many specialty vases will she need if she uses 25 pounds of roses?

10. Kendrick's Camaro can travel 124 miles on 4 gallons of gas. How many gallons of gas will he need to travel 300 miles?

11. Gloria needs 3 bags of fertilizer to fill 2 of her gardens. If she purchased 36 bags of fertilizer, how many gardens does she have?

12. Esco eats 2.5 bowls of dog food every 8 hours. How many bowls will he consume in one day?

13. If Sarah can run 100 yards in 12 seconds, how many yards can she run in 58 seconds?

14. Dr. Farris has performed 85 physicals this week, at this rate how many will she perform in one month?

15. Jimmy can lay 2 bags of mulch per hour. If his shift is 8 hours, how many bags of mulch will be lay?

16. If 8 candy bars weigh $1\frac{1}{9}$ lbs., what is the weight of 13 candy bars?

17. Ashanti can read 6 pages in 8 minutes. At this rate, how long will it take her to finish a book that has 245 pages?

18. The local supermarket sells apples $3.29/2lbs. At this rate, what is the cost of 9 lbs.?

19. For every pound a baby weighs, doctors recommend that they drink 2 oz of milk. If a baby drinks 21 oz of milk, how much does he weigh?

20. If one box of trash bags can last for 2 months. How many days can 3 boxes of trash bags last?

21. For every blue gummy bear in a bag there are 3 red gummy bears. If a bag contains 32 red gummy bears, how many blue gummy bears are there?

22. If David can type a 3-page paper in 2 hours, how long will it take him to type a 20-page paper?

23. House of JYL sells 100 items each week. At this rate, how many items will be sold in 4 months?

24. Brittany can bake 3 dozen cupcakes in $1\frac{1}{2}$ hours. At this rate, how many cupcakes can she bake in 9 hours?

25. If Stephanie can drive 250 miles in 3 hours, how far can she travel in 1.75 hours?

Answer Key

Practice 1

1. 525 miles
2. 600 words
3. 6.7 hours
4. 33 girls
5. 20 plants
6. 20 in
7. 1,050 miles
8. 30 stunts
9. 240 cattle
10. 1,500 words

Practice 2

1. $9\frac{5}{8}$ cups
2. $5\frac{1}{5}$ pages
3. $7\frac{6}{7}$ in
4. \$10.63
5. 4.64 bags
6. $2\frac{18}{25}$ in
7. $48\frac{1}{8}$ cartons
8. $1\frac{5}{16}$ lbs.
9. $3\frac{1}{3}$ in
10. $6\frac{3}{16}$ pies

Practice 3

1. 2,500 words
2. 48 pairs
3. 3 weeks
4. 4 weeks
5. 2700 words
6. 18 in
7. 320 bags
8. 64,800 words
9. 96 pencils
10. 4.2 months

Review

1. 8 hours
2. 135 cookies
3. $2\frac{1}{4}$ hours
4. 48 pairs
5. $14\frac{3}{10}$ pages
6. 19.2 hours
7. 63 pairs
8. $8\frac{16}{17}$ photoshoots
9. 8 vases
10. 9.7 gallons
11. 24 gardens
12. 7.5 bowls
13. $483\frac{1}{3}$ yards
14. 340 physicals
15. 16 bags
16. $1\frac{29}{36}$ lbs.
17. 326 minutes
18. \$14.81
19. 10.5 lbs.
20. 180 days
21. 10.7 blue bears
22. 13.3 hours
23. 1,600 items
24. 216 cupcakes
25. 145.8 miles

BLANK PAGE

Solve Real-World Problems Involving Percentages

You will learn:

How to calculate percentage increase

How to calculate percentage decrease

How to calculate tax, tip & discount

How to calculate percent markups

How to calculate percent markdowns

How to solve percent proportions

Study Tips

Read and study EVERY example problem.

Complete EVERY practice problem.

Check to make sure all answers are correct.

Go back to correct the questions you answered incorrectly.

If you don't receive at least an 80% on the review, go back and practice the topic.

Percentage Increase and Decrease

KISS IT

Step 1: Subtract the two values

Step 2: Divide the answer by the original value

Step 3: Convert the decimal to a percentage

The original value is the value that is changing

Example 1:

The normal price of denim jeans in House of JYL is $40, but today they're on sale for $25. What is the percent decrease in price?

Step 1: Subtract the two values $40 - $25 = $15

Step 2: Divide the answer by the original value 15 ÷ 40 = .375 b

Step 3: Convert the decimal to a percentage .375 × 100 = 37.5 %

Answer: Percent Decrease is 37.5%

Example 2:

During the summer movie tickets cost $5, but once Fall begins the normal price of $8.75 will be reinstated. What is the percent increase in price?

Step 1: Subtract the two values $8.75 - $5 = $3.75

Step 2: Divide the answer by the original value 3.75 ÷ 5 = .75

Step 3: Convert the decimal to a percentage .75 × 100 = 75

Answer: Percent Increase is 75%

Practice 1

1. Last year student enrollment at Exceptional Studies included 500 students. This year student enrollment includes 650 students, what is the percentage increase in enrollment?

2. Cakes are normally $25 at Quick Mart, but today the price has been reduced to $20. What is the percent decrease in price?

3. Alicia's usual hair appointment lasts for about 2 hours. Today her stylist kept her for 3 hours . What is the percent increase in time spent at the salon?

4. Last year Manny's favorite reality show had a total of 20 episodes. This year the number of episodes has increased to 28. What is the percent increase of episodes?

5. Devin has collected 500 pairs of sneakers over the last 5 years. Tomorrow he plans on donating 125 pairs to the local shelter. What will be the percentage decrease in the number of sneakers?

6. Last month Shank received 30 orders for custom sneakers. This month she received 20 more orders than last month. What is the percent increase in the number of orders?

7. A six-piece dinner at Bethune Grill is priced at $6.50. After using a coupon, Layla only paid $4.25 for her meal. What was the percent decrease in price?

8. When Sabrina was a child, a bag of chips cost 25 cents. Today the price for one bag is 45 cents. What is the percentage increase in price?

9. Last season Kevon completed 64 tackles. This season he completed 36 more tackles. What is the percent increase in tackles?

10. Last week Sheena cooked 9 meals for her family. This week she cooked 5 meals, what is the percent decrease in the number of meals she cooked?

Tax Tip & Discount

KISS IT

Step 1: Calculate the total

Step 2: Convert the percentage to a decimal

Step 3: Multiply the values

Example 3

Tori purchased a pair of jeans for $24.99, a blouse for $12.95, and a necklace for $6.99. If the sales tax rate is 6.5% in Florida, how much will she pay **in taxes?**

Step 1: Calculate the total $24.99 + 12.95 + 6.99 = 44.93$

Step 2: Convert % to decimal $6.5 \div 100 = .065$

Step 3: Multiply the values $44.93 \times .065 = 2.92$

Answer: Tori will pay $2.92 in taxes

Example 4

Tori purchased pair of jeans for $24.99, a blouse for $12.95, and a necklace for $6.99. If the sales tax rate is 6.5% in Florida, how much will she pay **at checkout?**

Step 1: Calculate the total $24.99 + 12.95 + 6.99 = 44.93$

Step 2: Convert % to decimal $6.5 \div 100 = .065$

Step 3: Multiply the values $44.93 \times .065 = 2.92$

Step 4: Add the values $44.93 + 2.92 = 47.85$

(Step 1 & Step 3)

Answer: Tori will pay $47.85 at checkout

Example 5

Tori purchased pair of jeans for $24.99, a blouse for $12.95, and a necklace for $6.99. The store is having a sale for 25% off the entire store. How much of a **discount** does she receive off her purchase?

Step 1: Calculate the total $24.99 + 12.95 + 6.99 = 44.93$

Step 2: Convert % to decimal $25 \div 100 = .250$

Step 3: Multiply the values $44.93 \times .25 = 11.23$

Answer: Tori will receive $11.23 off her purchase

Example 6

Tori purchased pair of jeans for $24.99, a blouse for $12.95, and a necklace for $6.99. The store is having a sale for 25% off the entire store. How much will she pay at checkout including her discount?

Step 1: Calculate the total $24.99 + 12.95 + 6.99 = 44.93$

Step 2: Convert % to decimal $25 \div 100 = .25$

Step 3: Multiply the values $44.93 \times .25 = 11.23$

Step 4: Subtract the values $44.93 - 11.23 = 33.7$

(Step 1 & Step 3)

Answer: Tori will pay $33.70

Example 7

Adrianne and her friends went out to dinner at the local steakhouse for their girls night. Her food cost $23.95, Michelle's food cost $42.50, and Gloria's food cost $18.95. If they decide to tip the waitress 15%, how much will they pay in gratuity?

Step 1: Calculate the total $23.95 + 42.50 + 18.95 = 85.40$

Step 2: Convert % to decimal $15 \div 100 = .15$

Step 3: Multiply the values $85.40 \times .15 = 12.81$

Answer: They will pay $12.81 in gratuity

Example 8

Adrianne and her friends went out to dinner at the local steakhouse for their girl's night. Her food cost $23.95, Michelle's food cost $42.50, and Gloria's food cost $18.95. If they decide to tip the waitress 15%, how much will they pay including the tip?

Step 1: Calculate the total $23.95 + 42.50 + 18.95 = 85.40$

Step 2: Convert % to decimal $15 \div 100 = .15$

Step 3: Multiply the values $85.40 \times 0.15 = 12.81$

Step 4: Add the values (Step 1 & Step 3) $85.40 + 12.81 = 98.21$

Practice 2

1. Bridget purchased a pair of sandals for $22.99, a belt for $14.99, a dress for $45.99, and a jumpsuit for $64.99. If the sales tax rate is 5%, how much will she **pay in taxes**?

2. Bridget purchased a pair of sandals for $22.99, a belt for $14.99, a dress for $45.99, and a jumpsuit for $64.99. If the sales tax rate is 5%, how much will she pay at **checkout**?

3. Bridget purchased a pair of sandals for $22.99, a belt for $14.99, a dress for $45.99, and a jumpsuit for $64.99. The store is having a sale for 30% off the entire store. How much of a **discount** does she receive off her purchase?

4. Bridget purchased a pair of sandals for $22.99, a belt for $14.99, a dress for $45.99, and a jumpsuit for $64.99. The store is having a sale for 30% off the entire store. How much will she pay at checkout including her discount?

5. Sabrina and her daughters decided to eat brunch today at Aunt Catfish's, and their bill came to a total of $75.69. If they agreed to tip the waitress 20%, how much will they pay in gratuity?

6. Sabrina and her daughters decided to eat brunch today at Aunt Catfish's, and their bill came to a total of $75.69. If they agreed to tip the waitress 20%, how much will they pay including the tip?

7. Shaunte spent $250 in Macy's, $125 in Dillard's, and $100 at GameStop for her son's birthday. If the sales tax rate is 8%, what it the total amount of money she spent including tax?

8. Teshia's date night included a trip to the movies, dinner, and drinks at the beach totaling $150. If she used coupons for 30% off each purchase, how much did she spend altogether?

9. Kendrick purchased new pots for $159.99. If the sales tax rate is 9.5%, how much did he pay in tax?

10. Lisa purchased a new computer for $699.99, virus protection for $129.99, and a wireless mouse for $19.99. If the store is having a sale for 20% off purchases, how much will she pay including the discount?

Percent Markups & Markdowns

KISS IT

STEP 1: Convert the percentage to a decimal

Step 2: Multiply the two values

Step 3: Add values (markup) or Subtract values (markdown)

Example 9

Jasmine purchased sundresses for $25 each wholesale. If she **marks up** the price by 65%, what will be the selling price?

Step 1: Convert the percent to decimal	$65 \div 100 = .65$
Step 2: Multiply the values	$25 \times .65 = 16.25$
Step 3: Add values	$25 + 16.25 = 41.25$

Answer: The selling price will be $41.25

Example 10

The normal price of denim jeans in House of JYL is $40, but the price has been **marked down** by 30%. What is the new selling price?

Step 1: Convert the percent to decimal	$30 \div 100 = .30$
Step 2: Multiply the values	$40 \times .30 = 12$
Step 3: Subtract values	$40 - 12 = 28$

Answer: The selling price is $28

Practice 3

1. The beauty store sells 24-inch bundles for $90 each. The price has been marked down 20% for the upcoming holiday sale. What is the new price?

2. Yesterday gasoline was priced at $3.05 per gallon. Today the price has been marked up by 15%. How much does gasoline cost today?

3. Smartphones are regularly priced at $300, but they've been marked down 10%. What is the new price?

4. Raekia's price for a headshot is normally $120, but the price was marked down 40% for a flash sale. What is the new price?

5. Alexis purchased jeans for $30 wholesale. She marked up the price 200% for resale in her boutique. What is the new selling price?

Percent Proportions

KISS IT

Step 1: Identify the percent, amount & whole

Step 2: Set up proportion

Step 3: Solve

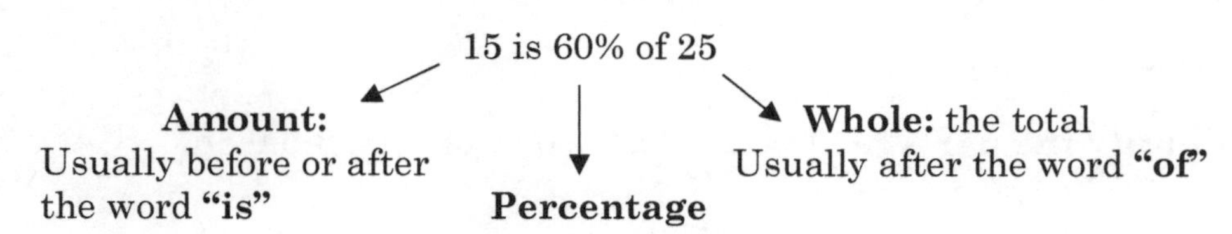

$$\frac{amount\ (is)}{whole\ (of)} = \frac{percent}{100}$$

Example 11

What is 25% of 150

Step 1: Identify the parts

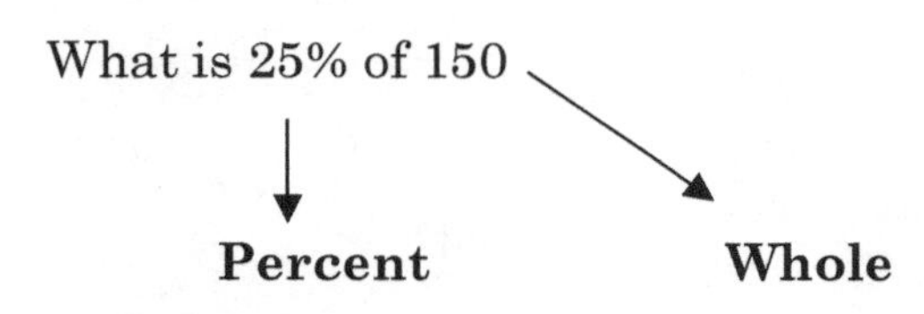

Step 2: Set up proportion

$$\frac{amount}{whole} = \frac{percent}{100}$$

$$\frac{amount}{150} = \frac{25}{100}$$

Step 3: Solve

$$\frac{amount}{150} = \frac{25}{100}$$

$150 \times 25 = 3750$

$3750 \div 100 = 37.5$

Example 12

10 is what percent of 50?

Step 1: Identify the parts

10 is what percent of 50

Amount **Whole**

Step 2: Set up proportion

$$\frac{amount}{whole} = \frac{percent}{100}$$

$$\frac{10}{50} = \frac{percent}{100}$$

Step 3: Solve

$$\frac{10}{50} = \frac{percent}{100}$$

$10 \times 100 = 1000$

$1000 \div 50 = 20$

Answer: 20%

Example 13

24 is 60% of what number?

Step 1: Identify the parts

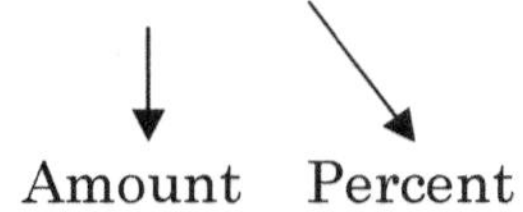

Amount Percent

Step 2: Set up proportion

$$\frac{amount}{whole} = \frac{percent}{100}$$

$$\frac{24}{whole} = \frac{60}{100}$$

Step 3: Solve

$$\frac{24}{whole} = \frac{60}{100}$$

$24 \times 100 = 2400$

$2400 \div 60 = 40$

Example 14

Jackson's yearly salary is $40,000. If 18% in taxes is taken out of his check, how much does he pay in taxes?

Step 1: Identify the parts $40,000: **whole** 18%: **percent**

Step 2: Set up proportion

$$\frac{amount}{whole} = \frac{percent}{100}$$

$$\frac{amount}{40{,}000} = \frac{18}{100}$$

Step 3: Solve

$$\frac{amount}{40{,}000} = \frac{18}{100}$$

$40{,}000 \times 18 = 720{,}000$

$720{,}000 \div 100 = 7200$

Answer: He pays $7,200 in taxes

Example 15

There are 28 students in Ms.Perriman's class, and 18 of them are boys. What percentage of her class do the boys represent?

Step 1: Identify the parts 28: **whole** 18: **amount**

Step 2: Set up proportion

$$\frac{amount}{whole} = \frac{percent}{100}$$

$$\frac{18}{28} = \frac{percent}{100}$$

Step 3: Solve

$$\frac{18}{28} \rightarrow \frac{percent}{100}$$

$$18 \times 100 = 1800$$

$$1800 \div 28 = 64.29$$

Answer: Boys represent 64.29% of her class

Practice 4

1. What is 25% of 60?

2. 15 is what percent of 45?

3. 20 is 75% of what number?

4. 12 is what percent of 50?

5. What is 45% of 270?

6. Mainland's football team won 10 out of 12 games this season. What percentage of games did they win?

7. Dresses represented 20% of sales at Lisa's boutique last month. If she sold a total of 150 items last month, how many of them were dresses?

8. Jeans are on sale for $20 along with an additional 25% off discount. What is the new price of the jeans?

9. 15% of the students in Stephanie's class failed the final exam. If she has a total of 30 students in her class, how many of them passed the exam?

10. There are 15 girls and 20 boys on the kickball team. What percentage of the team is represented by boys?

11. 62 is what percent of 200?

12. What is 18% of 48?

13. 30 is 25% of what number?

14. What is 29% of 150?

15. 25 is what percent of 475?

16. Of the graduating class, 52% of the students plan to attend college in state and the remaining plan to travel. If there are 200 students graduating, how many are planning to attend college out of state?

17. The local shelter houses 125 transients each night. If 75 of the transients are male, then what percentage of them are female?

18. If House of JYL has sold 80% of its new inventory, how many items are left if they originally stocked 150 new items?

19. Lori's wallet contains 4 credit cards, 2 debits cards,1 ID, and 1 passport. What percentage of the items do credit cards represent?

20. The local dance team has competed in 50 competitions this year, 45 of which they won. What percentage of their competitions did they win?

Review

1. Pizza at the movie costs $3.50, but on Tuesdays the price is reduced to $2.25. What is the percent decrease in price?

a. 56% **b.** 36% **c.** 34% **d.** 54%

2. Monday there were 20 girls trying out for the cheer team. By Wednesday there were 36 girls competing for a spot on the team. What is the percent increase in girls trying out?

a. 80% **b.** 8% **c.** 4% **d.** 44%

3. The local steakhouse automatically applies a 20% gratuity to parties of 6 or more. If the Baker's family dinner came to $360, how much was their total bill?

a. $72 **b.** $1.2 **c.** $432 **d.** $332

4. What is 85% of 200?

a. 170 **b.** 30 **c.** 425 **d.** 115

5. Destiny receives 25% off her purchase when she uses her employee discount at local stores. She picked up a pair of jeans for $79.99 and a blouse for $59.99. How much is her total bill including her discount?

a. $19.99 **b.** $14.99 **c.** $114.98 **d.** $104.99

6. What percent of 60 is 20?

a. 30% **b.** 33% **c.** 40% **d.** 300%

7. The local baseball team has won 80% of their games this season. If they've played 20 games so far, how many games have they won?

a. 16 **b.** 18 **c.** 25 **d.** 10

8. Keith purchased plane tickets for $105 each. If he purchased 3 tickets, and sales tax is 6% how much did he spend altogether on tickets?

a. 63 **b.** $18.90 **c.** $315 **d.** $334

9. Amber purchased dresses for $10 each and plans to resell them for 75% more than what she paid, what will be the new price?

a. $17.50 **b.** $13.33 **c.** $15 **d.** $10

10. David's favorite shoe store is having a huge sale for the upcoming holiday: 75% off the entire store. If the items that he picked up totaled $500, how much money will he receive off his purchase using the discount?

a. $125 **b.** $875 **c.** $375 **d.** $425

11. 60 is 50% of what number?

a. 90 **b.** 100 **c.** 120 **d.** 150

12. Natalie's last paycheck was $3,267.50 before taxes. If she pays 12% in federal taxes, how much is her paycheck after taxes?

a. $392.10 **b.** $3,659.60 **c.** $2,875.40 **d.** $3,267.50

13. Ashely paid $8.59 for breakfast, $12.45 for lunch, and $24.99 for dinner. Sales tax is 7.5%, how much did she pay in tax for breakfast and dinner?

a. $2.52 **b.** $3.45 **c.** $49.48 **d.** $36.10

14. Fur carpets have been marked down 25% from their regular price of $89.99. What is the new price after the percent markdown?

a. $64.99 **b.** $112.49 **c.** $22.50 **d.** $67.49

15. Alice's private car ride from the airport to the hotel costs $75. She decided to provide the driver with a 15% tip. How much was his tip?

a. $86.25 **b.** $60 **c.** $75 **d.** $11.25

16. What is 32.5% of 68?

a. 209.23 **b.** 22.1 **c.** 32.5 **d.** 221

17. Ms. Jefferson's original roster included 30 students. After the first two weeks, there were 21 students remaining. What is the percent decrease in the number of students on her roster?

a. 9% **b.** 3% **c.** 30% **d.** 42%

18. 38 is what percent of 125?

a. 30.4% **b.** 3.28% **c.** 47.5% **d.** 32.8%

19. Last week's rival football game attracted a crowd of 850. If 60% of the crowd consisted of males, how many females attended the game?

a. 40 **b.** 790 **c.** 340 **d.** 510

20. In the last election, 42% of the voters identified as Democrats, 40% identified as Republicans, and the remaining identified themselves as Independents. What percentage of the voters identified themselves as Independent?

a. 18% **b.** 42% **c.** 40% **d.** 82%

Answer Key

Practice 1

1. 30%

2. 20%

3. 50%

4. 40%

5. 25%

6. 67%

7. 35%

8. 80%

9. 56%

10. 44%

Practice 2

1. $7.45

2. $156.41

3. $44.69

4. $104.27

5. $15.14

6. $90.83

7. $513

8. $105

9. $15.20

10. $679.98

Practice 3

1. $72

2. $3.51

3. $270

4. $72

5. $90

Practice 4

1. 15

2. 33%

3. 26.7

4. 24%

5. 121.5

6. 83%

7. 30

8. $15

9. 25

10. 57%

11. 31%

12. 8.64

13. 120

14. 43.5

15. 5.26%

16. 96

17. 40%

18. 30

19. 50%

20. 90%

Review

1. B

2. A

3. C

4. A

5. D

6. B

7. A

8. D

9. A

10. C

11. C

12. C

13. A

14. D

15. D

16. B

17. C

18. A

19. C

20. A

Cluster 2 Review

You should know:

☐ How to write the ratio between two quantities

☐ How to identify equivalent ratios

☐ How to solve for missing values given a ratio

☐ How to solve proportions with whole numbers

☐ How to solve proportions with fractions and decimals

☐ How to solve proportions containing values with different units

☐ How to calculate percentage increase

☐ How to calculate percentage decrease

☐ How to calculate tax, tip & discount

☐ How to calculate percent markups

☐ How to calculate percent markdowns

☐ How to solve percent proportions

Cluster 2 Review

1. What percent of 120 is 75?

a. 16% **b.** 62.5% **c.** 160% **d.** 625%

2. Alexis can type $3\frac{1}{4}$ pages in 2 hours. At this rate, how many pages can she type in 6 hours?

a. $6\frac{1}{2}$ **b.** $9\frac{3}{4}$ **c.** $3\frac{9}{13}$ **d.** $18\frac{1}{4}$

3. The ratio of boys to girls in the school choir is 1 to 3. If there is a total of 21 girls in the choir, how many boys are there?

a. 21 **b.** 63 **c.** 7 **d.** 8

4. The average cost of bottled water increased from $2.00 to $2.29. What is the percent increase in price?

a. 14.5% **b.** 12.7% **c.** 14% **d.** 13%

5. The softball team won 75% of its games so far this season. If they've won 15 games so far, how many games have they played overall?

a. 11 **b.** 500 **c.** 20 **d.** 15

6. Darryl can drive 65 miles on 2 gallons of gas. If he fills his tank with 10 gallons, how far can he travel?

a. .31 **b.** 3.25 **c.** 130 **d.** 325

7. A bag includes 6 blue marbles, 4 red marbles, 7 yellow marbles, and 5 green marbles. What is the ratio of blue and red marbles to the total number of marbles in the bag?

a. $\frac{10}{21}$ **b.** $\frac{5}{11}$ **c.** $\frac{3}{2}$ **d.** $\frac{11}{5}$

8. Brian's dry-cleaning bill came to a total of $175 before taxes. If the local tax rate is 5%, how much was his bill including tax?

a. $183.75 **b.** $9 **c.** $8.75 **d.** $184

9. The ratio of boys to girls in 6th grade is 2:3. If the ratio of boys to girls is the same in 8th grade, identify an equivalent ratio.

a. $\frac{30}{45}$ **b.** $\frac{3}{2}$ **c.** $\frac{45}{30}$ **d.** $\frac{4}{9}$

10. Gina purchased a pair of denim shorts that were on sale for $25. If the original price was $40, what was the percent decrease in price?

a. 37.5% **b.** 16% **c.** 40% **d.** 60%

11. Ashley purchased 8 pairs of shoes in two weeks. If she continues to shop at this rate, how many shoes will she purchase in 3 months?

a. 16 **b.** 9 **c.** 32 **d.** 48

12. The ratio of altos to sopranos in the church choir is 1 to 3. If there are 9 sopranos how many altos are there?

a. 21 **b.** 3 **c.** 7 **d.** 27

13. Aaliyah always tips the waitress 15%. If her total bill came to $23.49, how much of a tip will she leave?

a. $27.01 **b.** $23.49 **c.** $3.52 **d.** $3.50

14. There are 60 football players, 35 basketball players, and 28 baseball players visiting the local college for a sports camp. What is the ratio of basketball players to football players?

a. 25:60 **b.** 7:12 **c.** 12:7 **d.** 60:35

15. If Stephanie's nephew can eat $1\frac{1}{2}$ pizzas in 2 hours, how many can he eat in 5 hours?

a. $\frac{3}{5}$ **b.** 3 **c.** $3\frac{3}{4}$ **d.** 4

16. Alicia can bake 10 cakes in one day. At this rate, how many cakes can she bake in 2 weeks?

a. 20 **b.** 70 **c.** 100 **d.** 140

17. Employees at Shoe World receive a discount of 20% off very purchase. If one employee spends $75 and another employee spends $100, how much money will they receive off their purchases together?

a. $35 **b.** $15 **c.** $80 **d.** $5

18. The ratio of men to women at Brandon's gym is 1:4. If the local gym has an equivalent ratio, identify a possible ratio from the selection below.

a. 4 to 7 **b.** 6 to 24 **c.** 24 to 6 **d.** 7 to 4

19. Leslie completes 12 pages of notes every hour she's at work. At this rate, how many can she complete in one 12-hour shift?

a. 144 **b.** 12 **c.** 96 **d.** 24

20. Ashley's gross pay last year was $46,986. If she paid 15% in taxes, how much was her take home pay?

a. $39,938 **b.** $54,034 **c.** $7,048 **d.** $46,986

Answer Key

1. B

2. B

3. C

4. A

5. C

6. D

7. B

8. A

9. A

10. A

11. D

12. B

13. C

14. B

15. C

16. D

17. A

18. B

19. A

20. A

Cluster 2 Review Break Down

Solve Real-World Problems Involving Ratios and Rate of Change

Question 3 ☐

Question 7 ☐

Question 9 ☐

Question 12 ☐

Question 14 ☐

Question 18 ☐

Question 19 ☐

Solve Real-World Problems Involving Proportions

Question 2 ☐

Question 6 ☐

Question 11 ☐

Question 15 ☐

Question 16 ☐

Solve Real-World Problems Involving Percentages

Question 1 ☐

Question 4 ☐

Question 5 ☐

Question 8 ☐

Question 10 ☐

Question 13 ☐

Question 17 ☐

Question 20 ☐

BLANK PAGE

Cluster 3

Explain the Relationship Between Two Variables

- How to identify the graph that best illustrates a scenario
- How to calculate statistics given a graph
- How to calculate percentages given a graph
- How to reach valid conclusions given a graph

Evaluate Information in Tables Charts and Graphs Using Statistics

- How to calculate mean
- How to calculate median
- How to calculate mode
- How to calculate range
- How to classify the shape of a distribution

Interpret Relevant Information from Tables Charts and Graphs

- How to identify the graph that best illustrates a scenario
- How to calculate statistics given a graph
- How to calculate percentages given a graph
- How to reach valid conclusions given a graph

Calculate Geometric Quantities

- How to calculate the perimeter of a figure
- How to calculate the area of a figure
- How to solve for the missing dimension of a figure

BLANK PAGE

Explain the Relationship Between Two Variables

You will learn:

How to identify independent variables

How to identify dependent variables

How to identify correlation given a graph

How to identify correlation given a scenario

Study Tips

Read and study EVERY example problem.

Complete EVERY practice problem.

Check to make sure all answers are correct.

Go back to correct the questions you answered incorrectly.

If you don't receive at least an 80% on the review, go back and practice the topic.

INDEPENDENT VARIABLE (CAUSE)

CAN STAND ON ITS OWN CAN CHANGE ON ITS OWN

CAN BE CHANGED BY EXPERIMENTER

ISN'T CHANGED BY OTHER VARIABLES BEING MEASURED

"CAUSES" A CHANGE IN ANOTHER VARIABLE

INPUT X-VALUE

Common Examples

Time **Age**

DEPENDENT VARIABLE (EFFECT)

Value depends on another Value being measured

Value being "affected"

Output Y-value

Common Examples

Height **Weight**

***Tip**

The independent variable is what you can change, and the dependent variable is what changes because of that.

Example 1

Identify the independent and dependent variables in the given scenario

A teacher examines the impact of time spent studying on test scores

KEYWORD: time is ALWAYS the independent variable

KEYWORD: the independent variable always follows the phrase **"impact of"**

Independent variable: time **Dependent variable: test score**

Practice 1

1.

Hours Practicing	Number of wins
2	0
4	2
6	5
8	7
10	9

Independent Variable:

Dependent Variable:

2. A school district studies the impact of teacher pay on job satisfaction.

Independent Variable:

Dependent Variable:

3. A restaurant owner studies the effect of price on the number of burgers sold.

Independent Variable:

Dependent Variable:

4. A football player thinks he performs better when it's raining.

Independent Variable:

Dependent Variable:

5. My mother believes that if your shower before bed, you will sleep longer.

Independent Variable:

Dependent Variable:

6. The more time people spend on their phones, the less able they are to communicate effectively.

Independent Variable:

Dependent Variable:

7. Exercising regularly increases happiness.

Independent Variable:

Dependent Variable:

8. Less sleep impacts aggressiveness.

Independent Variable:

Dependent Variable:

Positive Correlation: a relationship between two variables in which values move in the same direction together. If one variable increases, then the other variable increases. If one variable decreases, then the other variable decreases.

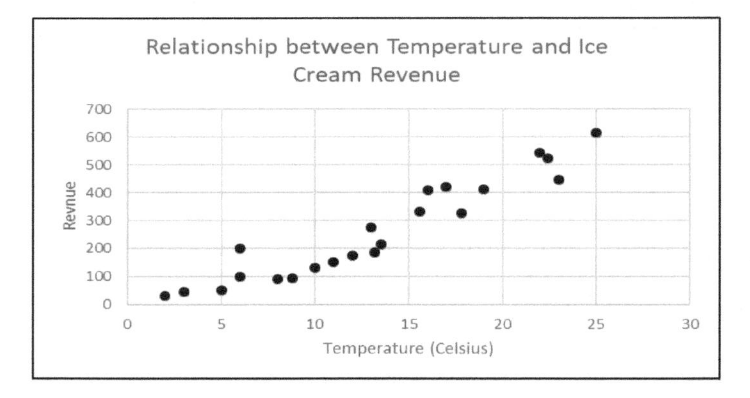

Negative Correlation: a relationship between two variables in which values move in the opposite direction together. If one variable increases, then the other decreases. If one variable decreases, then the other increases.

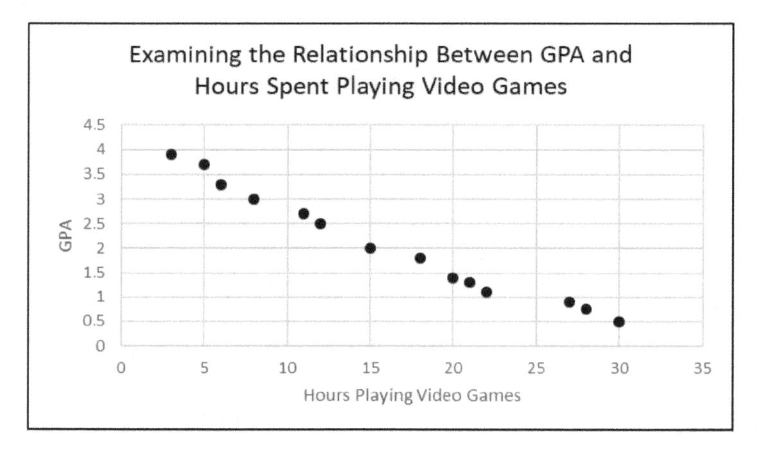

Zero Correlation: this type of correlation indicates that no relationship exists between two variables.

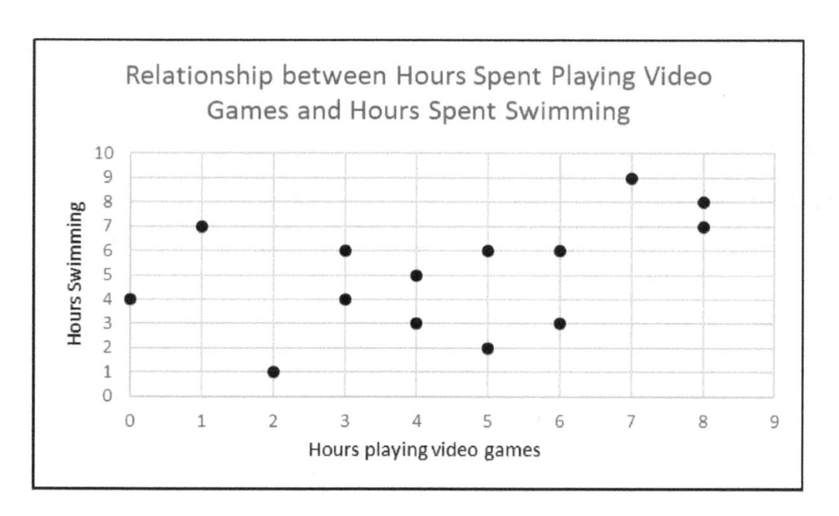

Example 2

Identify the type of correlation represented in the given scenario.

As the temperature increased in Florida, ice cream sales skyrocketed as well.

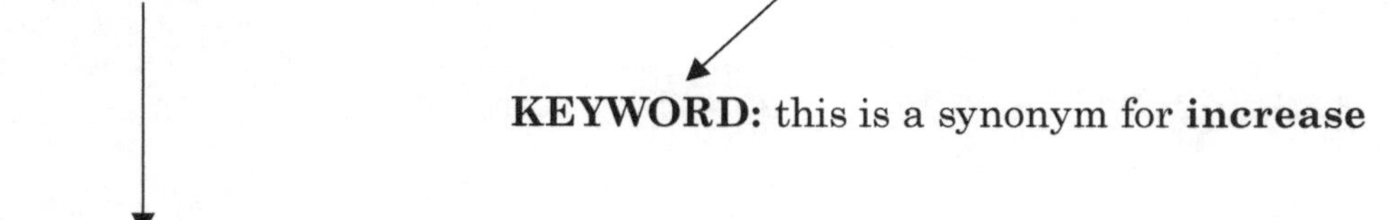

KEYWORD: this is a synonym for **increase**

KEYWORD

Step 1:Identify variables	**Step 2: Indicate the direction of each variable**
Variable 1: Temperature	Variable 1: Increased
Variable 2: Ice cream sales	Variable 2: Increased

Both variables moved in the same direction indicating a **positive correlation.**

Example 3

Identify the type of correlation represented in the given scenario.

Facebook's stock increased last quarter after they experienced a drop in the number of fraudulent accounts.

KEYWORD: this is a synonym for decrease

KEYWORD

Step 1:Identify variables variable	**Step 2: Indicate the direction of each**
Variable 1: Stock	Variable 1: Increased
Variable 2: Number of accounts	Variable 2: Decreased

The variables moved in different directions indicating a **negative correlation**

Practice 2

Identify the type of correlation represented in the given scenarios.

1.

Hours Practicing	Number of wins
2	0
4	2
6	5
8	7
10	9

a. positive correlation

b. negative correlation

c. zero correlation

2.

Calorie Intake	Weight Loss
1200	-10
1500	-8
1900	-5
2200	-3
2500	-2
3000	-1

a. positive correlation

b. negative correlation

c. zero correlation

3. The more money Brittany makes, the more money she spends on shoes.

a. positive correlation

b. negative correlation

c. zero correlation

4. The less time parents spend with their children, the more likely they are to exhibit behavioral issues.

a. positive correlation

b. negative correlation

c. zero correlation

5. The amount of water consumed and IQ level

a. positive correlation

b. negative correlation

c. zero correlation

Review

1. Distance traveled and the amount of gasoline in your gas tank

a. positive correlation **b.** negative correlation **c.** zero correlation

2. The more Tyler works, the less he begins to like his job

a. positive correlation **b.** negative correlation **c.** zero correlation

3. Distance traveled and the amount of gasoline in your gas tank

Independent Variable:

Dependent Variable:

4. Students studied the effect of their behavior on teacher attitude

Independent Variable:

Dependent Variable:

5. Define the following terms:

- positive correlation:
- negative correlation:
- zero correlation:

6. The more it rains, the more car accidents occur

a. positive correlation **b.** negative correlation **c.** zero correlation

7. Height and political affiliation

a. positive correlation **b.** negative correlation **c.** zero correlation

8. A study measuring which brand of laundry detergent removes stains the best

Independent Variable:

Dependent Variable:

9. Boys grow taller as they age

Independent Variable:

Dependent Variable:

10. Student enrollment and number of teachers on staff

Independent Variable:

Dependent Variable:

11. Plant heights vary depending on the amount of fertilizer they receive.

Independent Variable:

Dependent Variable:

12. Getting a good night's rest helps students to focus better in school

Independent Variable:

Dependent Variable:

13. IQ and weight

a. positive correlation **b.** negative correlation **c.** zero correlation

14. Number of traffic violations and monthly insurance premium

a. positive correlation **b.** negative correlation **c.** zero correlation

15. Level of education and communication skills

a. positive correlation **b.** negative correlation **c.** zero correlation

16. Salary and debt

a. positive correlation **b.** negative correlation **c.** zero correlation

17. Number of hours studying and test score

a. positive correlation **b.** negative correlation **c.** zero correlation

18. Spending time with family decreases stress

Independent Variable:

Dependent Variable:

19. Larger family size increase the likelihood of depression

Independent Variable:

Dependent Variable:

20. The more students read, the more they participate in class.

Independent Variable:

Dependent Variable:

Answer Key

Practice 1

1. Independent: Hours of practice — Dependent: Number of wins

2. Independent: Teacher pay — Dependent: Job satisfaction

3. Independent: Price — Dependent: Number of burgers sold

4. Independent: Rain — Dependent: Performance

5. Independent: Shower before bed — Dependent: Sleep

6. Independent: Time on social media — Dependent: Effectively communicating

7. Independent: Exercise — Dependent: Happiness

8. Independent: Amount of sleep — Dependent: Aggressiveness

Practice 2

1. A **2. B** **3. A** **4. B** **5. C**

Review

1. B **2. B**

3. Independent: Amount of gas — Dependent: Distance traveled

4. Independent: Student behavior — Dependent: Teacher attitude

5. a. a relationship in which variable change in the same direction

b. a relationship in which variables change in opposite directions

c. no relationship exists between variables

6. A **7. C**

8. Independent: Brand of detergent — Dependent: Stain removal

9. Independent: Age — Dependent: Height

10. Independent: Student enrollment — Dependent: Number of teachers

11. Independent: Amount of fertilizer — Dependent: Plant height

12. Independent: Amount of rest — Dependent: Ability to focus

13. C **14. A** **15. A** **16. B** **17. A**

18. Independent: Time with family — Dependent: Stress level

19. Independent: Family Size — Dependent: Likelihood of depression

20. Independent: Time spent reading — Dependent: Class participation

BLANK PAGE

Evaluate Information in Tables Charts and Graphs Using Statistics

You will learn:

How to calculate mean

How to calculate median

How to calculate mode

How to calculate range

How to classify the shape of a distribution

Study Tips

Read and study EVERY example problem.

Complete EVERY practice problem.

Check to make sure all answers are correct.

Go back to correct the questions you answered incorrectly.

If you don't receive at least an 80% on the review, go back and practice the topic.

Mean

The mean is the average of a data set.

KISS IT!

Step 1: Add the numbers in the data set

Step 2: Divide the sum by the total numbers in the data set

Example 1

Given the data set {2,6,8,4,6,10} calculate the mean.

Step 1: 2 + 6 + 8 + 4 + 6 + 10 = 36

Step 2: 36 ÷ **6** = 6

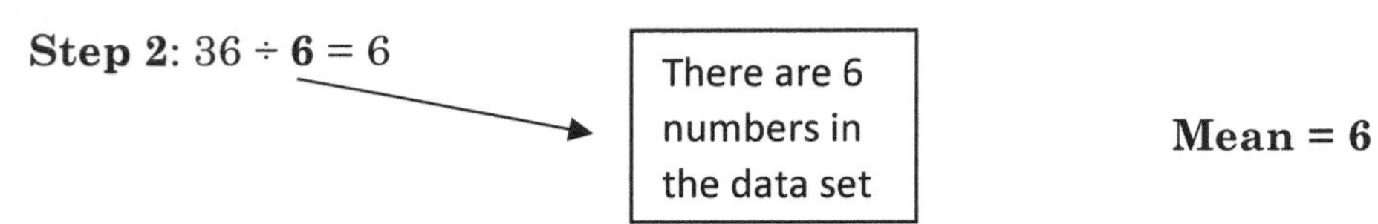

Mean = 6

Example 2

Given the data set, calculate the average.

Student Test Scores
26
34
32
23
17
29
19

Step 1: 26 + 34 + 32 + 23 + 17 + 29 +19 = 180

Step 2: 180 ÷ 7 = 25.7

There are 7 numbers in the data set

Mean = 25.7

Median

The median is the value located in the middle of a data set.

KISS IT!

Step 1: Place the data set in order from least to greatest

Step 2: If the data set is odd, then the median is the middle number

If the data set is even, the median is the average of the two middle numbers

Example 3

Given the data set {2,6,8,4,6,10} calculate the median.

Step 1: 2, 4, 6, 6, 8, 10

Step 2: This is an **even** data set (6 numbers)

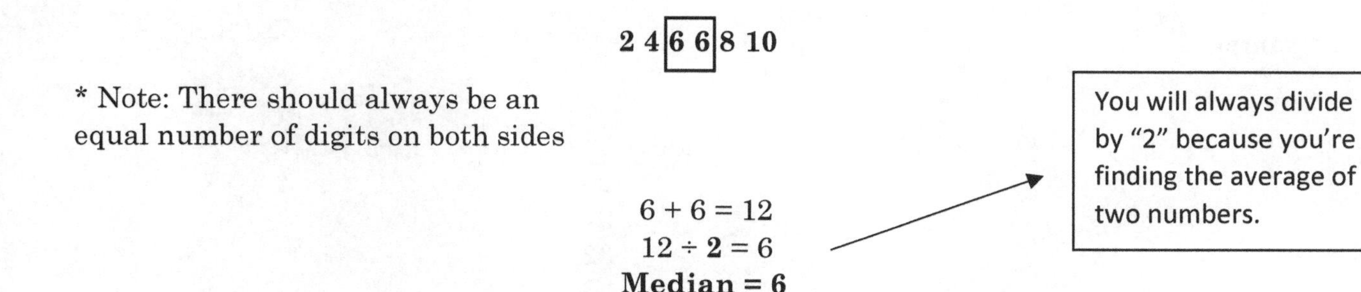

Example 4

Given the data set, calculate the median.

Student Test Scores
26
34
32
23
17
29
19

Step 1: 17 19 23 26 29 32 34

Step 2: This is an **odd** data set (7 numbers)

17 19 23 [26] 29 32 34

* Note: There should always be an equal number of digits on both sides

Median = 26

Mode

The mode is the number that is repeated the most in a data set.

KISS IT!

Step 1: Identify the number that is repeated the most

Example 5

Given the data set {2,6,8,4,6,10} calculate the mode.

2 [6] 8 4 [6] 10

Mode = 6

Example 6

Given the data set {6, 34, 67, 6, 23, 67, 8} calculate the mode.

[6] 34 [67] [6] 23 [67] 8

*Note: There are two numbers that occur the most, so therefore there will be two modes.

Mode = 6 & 67

Example 7

Given the data set, identify the mode.

Student Test Scores
26
34
32
23
17
29
19

* Notice that none of the data values in this data set repeat.

When this occurs, there is **no mode**.

Mode = There is no mode in this data set

<u>Range</u>

The **range** is the difference between the least and greatest values in a data set

KISS IT!

Step 1:Arrange the numbers from least to greatest

Step 2. Subtract the first and last numbers

Example 8

Given the data set {2,6,8,4,6,10} calculate the range.

Step 1: 2, 4, 6, 6, 8, 10

Step 2: 10 – 2 = 8

Range = 8

Example 9

Given the data set, calculate the range.

Student Test Scores
26
34
32
23
17
29
19

Step 1: 17 19 23 26 29 32 34

Step 2: 34 – 17 = 17

Range = 17

Practice 1

1. {6, 5, 48, 32, 15, 20, 5}

Mean: ____________

Median: ___________

Mode: _____________

Range: _____________

2. {14, 48, 65, 45, 19}

Mean: ____________

Median: ___________

Mode: _____________

Range: _____________

3. {3.2, 4.6, 3.2, 0, 5, 4.6}

Mean: ____________

Median: ___________

Mode: _____________

Range: _____________

4.

Grade level	Number of Athletes
8th	365
9th	95
10th	229
11th	401
12th	63

Mean: ____________

Median: ___________

Mode: _____________

Range: _____________

5.

Ice Cream Flavors	Sales
Vanilla	$252
Chocolate	$498
Strawberry	$251
Cookies & Cream	$380

Mean: _____________

Median: ___________

Mode: _____________

Range: ______________

6.

Month	Average Temperature
June	94
July	92
August	94
September	82
October	75
November	72

Which of the following is true?

a. Mean > mode
b. Median < Range
c. Mode = Mean
d. Range < Mean

7.

GPA
2.75
3.62
3.98
2.56
1.8
2.0

What conclusion can be drawn from the data set?

a. The average gpa is 2.8
b. This student's gpa range is 2.5
c. The most common gpa is 2.0
d. The middle gpa is 2.5

Graphs

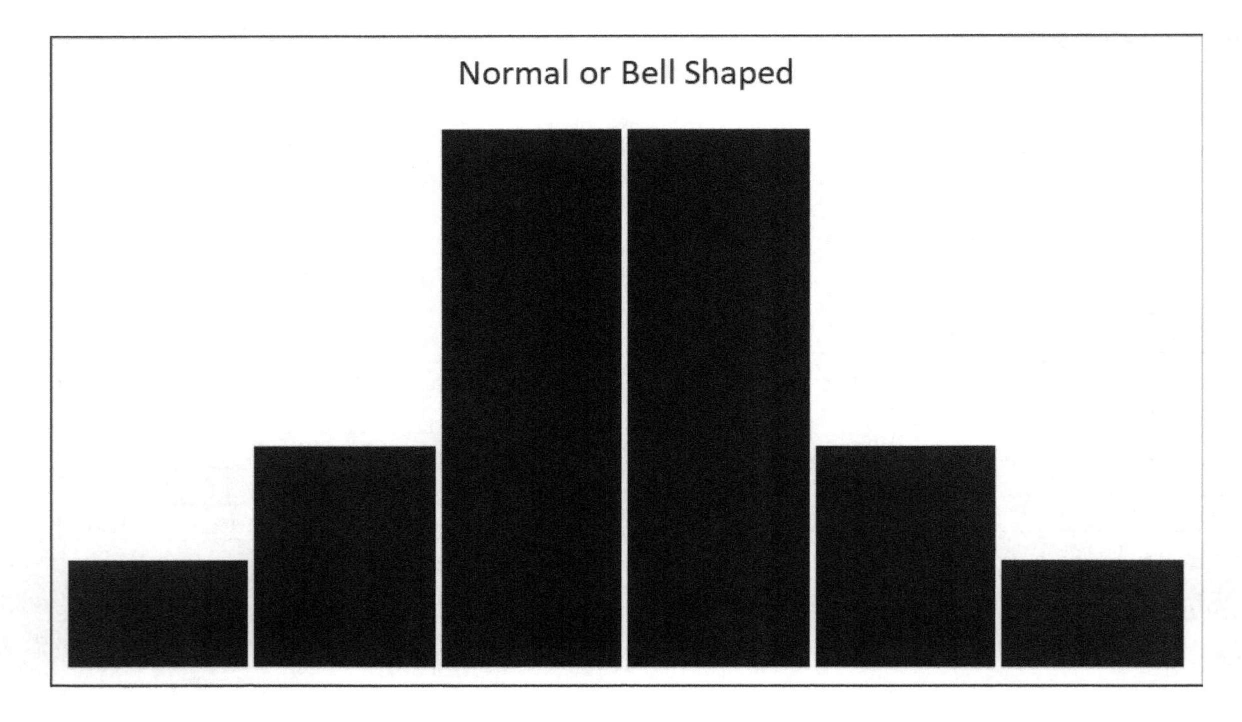

MEAN = MEDIAN

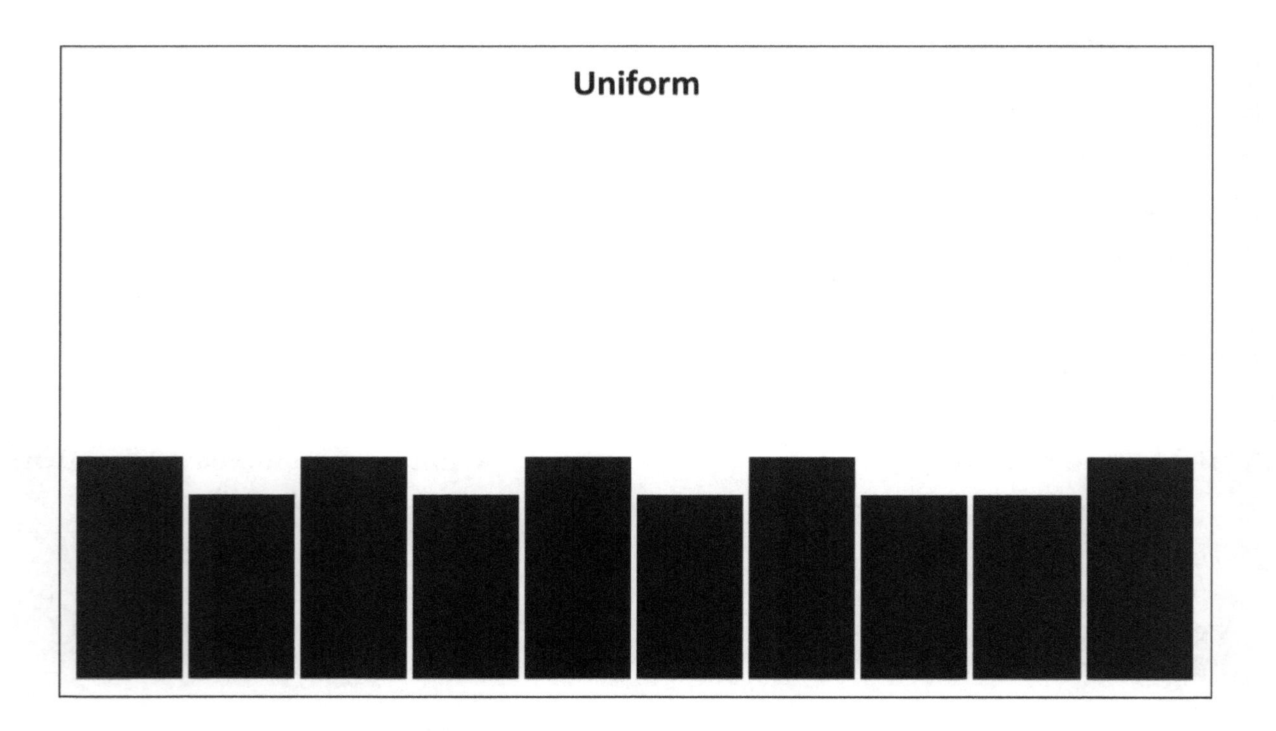

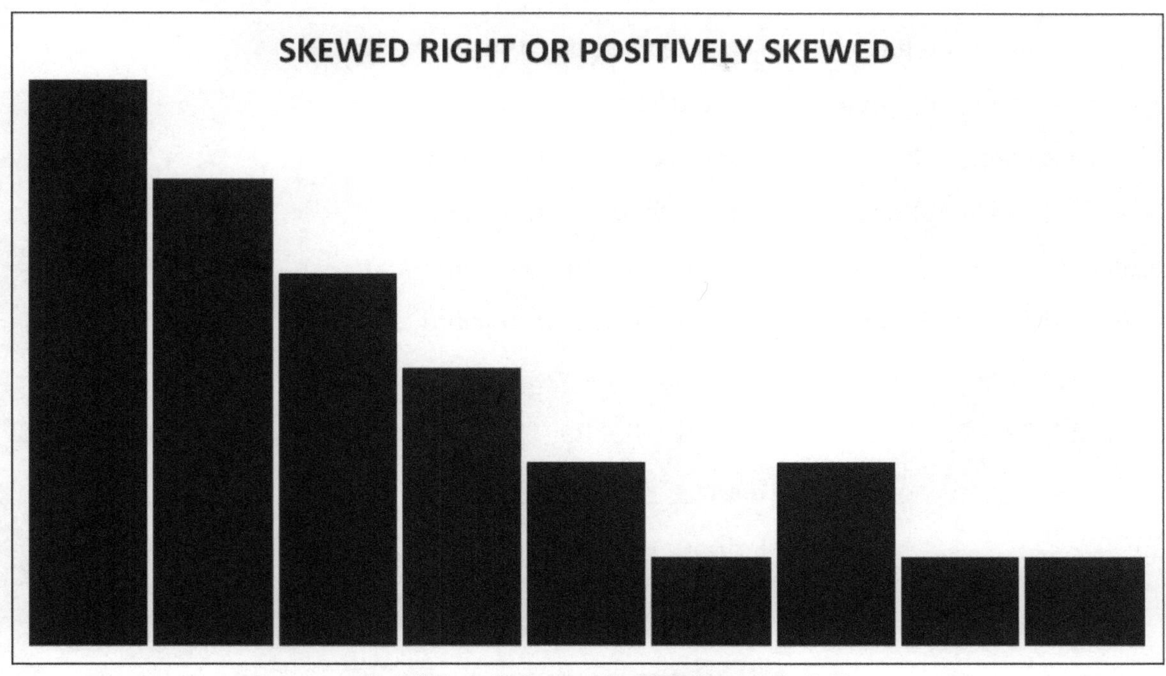

MEAN > MEDIAN

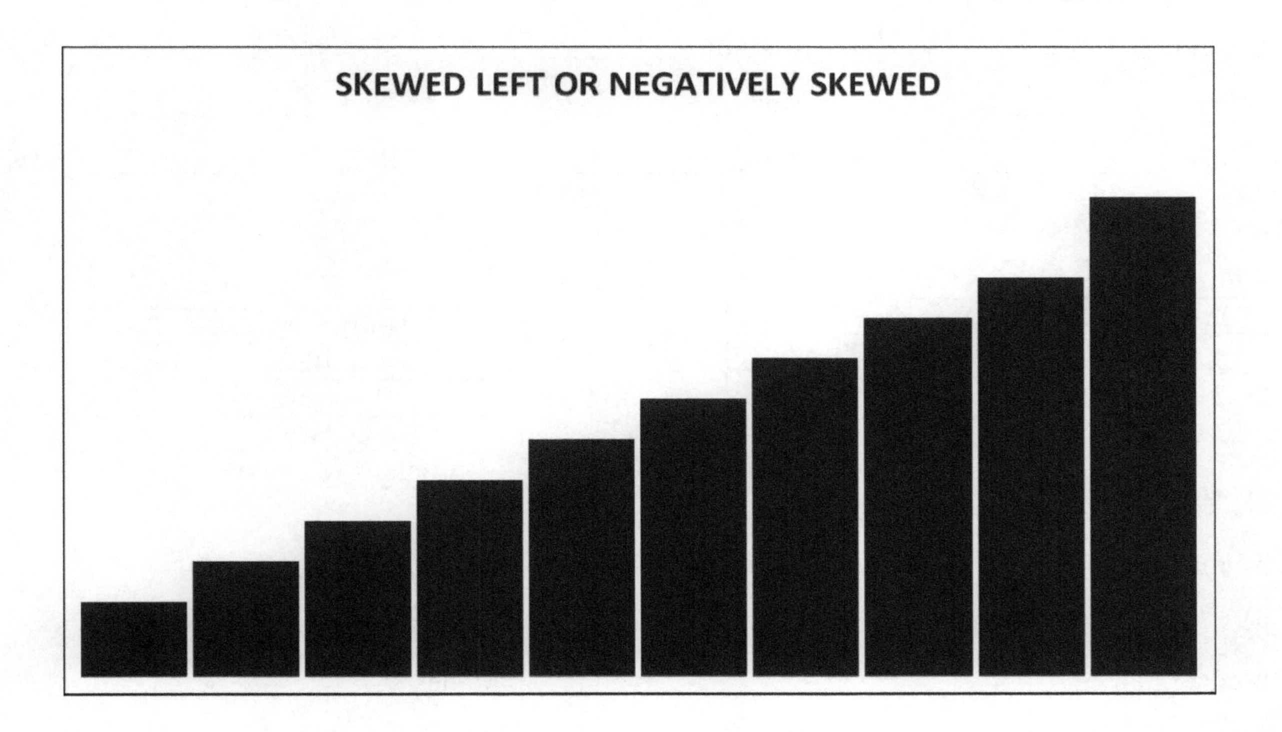

MEAN < MEDIAN

Review

1. Which graph shape has a mean that is greater than the median?

a. positive **b.** negative **c.** uniform **d.** normal

2. Which graph shape has a mean that is equal to the median?

a. positive **b.** negative **c.** uniform **d.** normal

3. Which graph shape has a mean that is less than the median?

a. positive **b.** negative **c.** uniform **d.** normal

4. {22, 47, 20, 22, 45, 46, 9}

Mean: ___________

Median: ___________

Mode: ___________

Range: ___________

5. {24, 10, 35, 39, 19}

Mean: ___________

Median: ___________

Mode: ___________

Range: ___________

6. {6.3, 7.6, 4.1, 3, 5.9, 4.6}

Mean: ___________

Median: ___________

Mode: ___________

Range: ___________

7. {62, 54, 48, 62, 15, 29, 25}

Mean: ___________

Median: ___________

Mode: ___________

Range: ___________

8. {17, 38, 61, 17, 20}

Mean: ___________

Median: ___________

Mode: ___________

Range: ___________

9. {32, 46, 32, 0, 5, 96}

Mean: ___________

Median: ___________

Mode: ___________

Range: ___________

10.

Classification	Nursing Majors
Freshman	105
Sophomore	89
Junior	65
Senior	32
Graduate Student	20

Mean: ___________

Median: ___________

Mode: ___________

Range: ___________

11.

Milk Shake Flavors	Sales
Vanilla	$144
Chocolate	$78
Strawberry	$100
Cookies & Cream	$225

Mean: ___________

Median: ___________

Mode: ___________

Range: ___________

12.

Month	Inches of Rain
June	24
July	23
August	18
September	15
October	8
November	6

Which of the following is true?

a. Mean > mode
b. Median < Range
c. Mode = Mean
d. Range < Mean

13.

GPA
3.89
1.34
4.21
2.59
1.80
2.03

What conclusion can be drawn from the data set?

a. The average gpa is 2.65
b. This student's gpa range is 2.87
c. The most common gpa is 4.0
d. The median gpa is 2.41

14. {2.4, 4.5, 4.2, 3.9, 2.4}

Mean: ____________

Median: ___________

Mode: _____________

Range: _____________

15. {21, 18, 45, 59, 13}

Mean: ____________

Median: ___________

Mode: _____________

Range: _____________

16. {9.3, 2.6, 4.3, 3, 2.6, 4.1}

Mean: _____________

Median: ____________

Mode: _____________

Range: ______________

17. {23, 27, 89, 50, 23, 27, 25}

Mean: ____________

Median: ___________

Mode: _____________

Range: _____________

18. {71, 83, 16, 71, 2}

Mean: ____________

Median: ___________

Mode: _____________

Range: _____________

19. {323, 64, 23, 0, 5, 69}

Mean: _____________

Median: ____________

Mode: _____________

Range: ______________

20. {6, 5, 4, 6, 1, 2, 2}

Mean: ____________

Median: ___________

Mode: _____________

Range: _____________

Answer Key

Practice 1

1. Mean: 18.7	Median: 15	Mode: 5	Range: 43
2. Mean: 38.2	Median: 45	Mode: no mode	Range: 51
3. Mean: 3.4	Median: 3.9	Mode: 3.2 & 4.6	Range: 5
4. Mean: 230.6	Median: 229	Mode: no mode	Range: 338
5. Mean: 345.3	Median: 316	Mode: no mode	Range: 247

6. D

7. A

Review

1. A	**2. D**	**3. B**	
4. Mean: 30.1	Median: 22	Mode: 22	Range: 38
5. Mean: 25.4	Median: 24	Mode: no mode	Range: 29
6. Mean: 5.3	Median: 5.3	Mode: no mode	Range: 4.6
7. Mean: 42.1	Median: 48	Mode: 62	Range: 47
8. Mean: 30.6	Median: 20	Mode: 17	Range: 44
9. Mean: 35.2	Median: 32	Mode: 32	Range: 96
10. Mean: 62.2	Median: 65	Mode: no mode	Range: 85
11. Mean: 136.75	Median: 122	Mode: no mode	Range: 147

12. B

13. B

14. Mean: 3.5	Median: 3.9	Mode: 2.4	Range: 2.1
15. Mean: 31.2	Median: 21	Mode: no mode	Range: 46
16. Mean: 4.3	Median: 3.6	Mode: 2.6	Range: 6.7
17. Mean: 37.7	Median: 27	Mode: 23, 27	Range: 66
18. Mean: 48.6	Median: 71	Median: 71	Range: 81
19. Mean: 80.7	Median: 43.5	Median: no mode	Range: 323
20. Mean: 3.7	Median: 4	Median: 2, 6	Range: 5

Interpret Relevant Information from Tables Charts and Graphs

You will learn:

How to identify the graph that best illustrates a scenario

How to calculate statistics given a graph

How to calculate percentages given a graph

How to reach valid conclusions given a graph

Study Tips

Read and study EVERY example problem.

Complete EVERY practice problem.

Check to make sure all answers are correct.

Go back to correct the questions you answered incorrectly.

If you don't receive at least an 80% on the review, go back and practice the topic.

Types of Graphs

Bar Graph

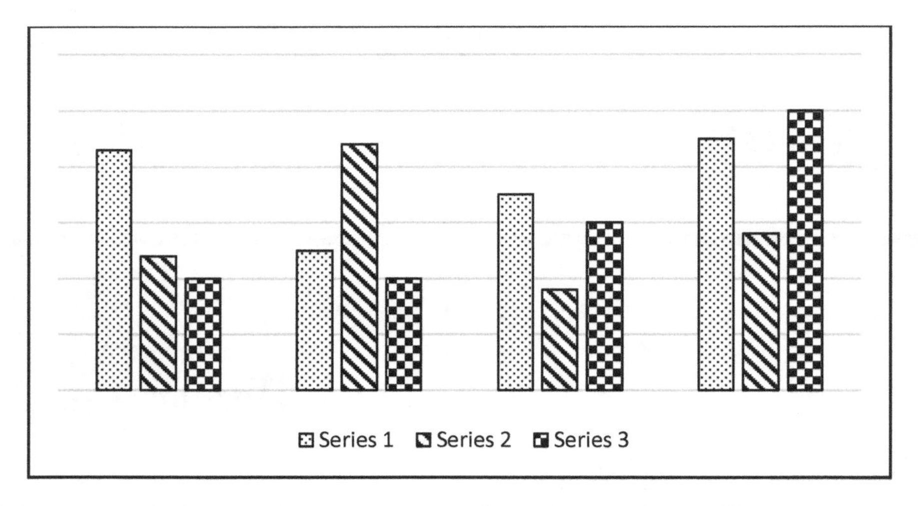

A bar graph is used to compare data between different groups

When to use

- To make comparisons between different variables
- To track changes over time

Histogram

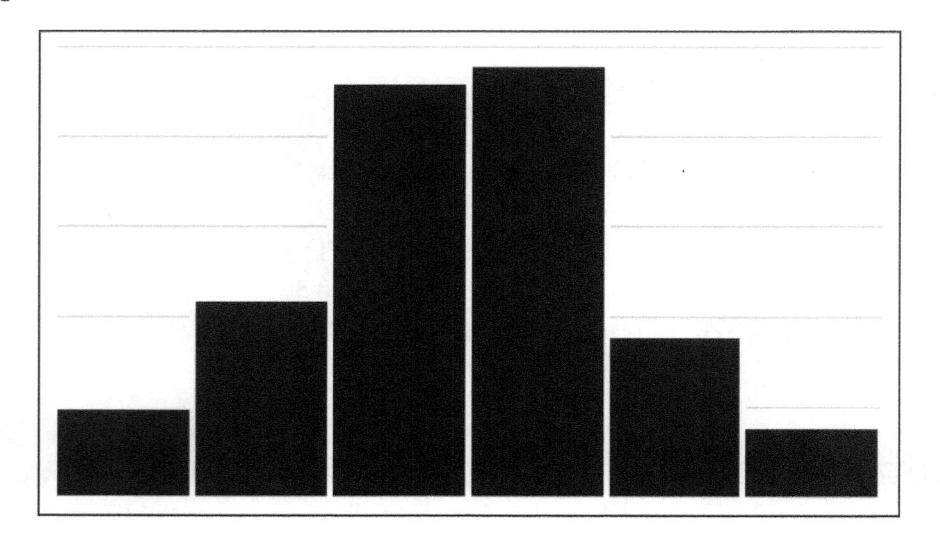

A histogram is used to display the frequency of given variables

When to use

- To show the shape of a distribution
- To organize a large number of variables

KISS IT!

Step 1: Identify what the question is asking

Step 2: Use the proper subskills to execute

Example 1

Based on the graph, what is the average number of scholarships awarded?

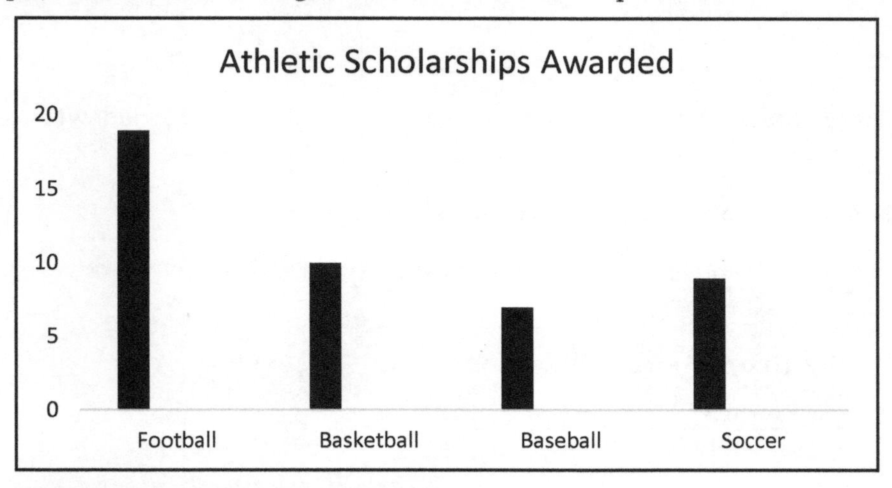

Step 1: Identify what the question is asking

Based on the graph, what is the **average** number of scholarships awarded?

Step 2: Use the proper subskills to execute

Average is the **same** as the **mean**

1. Add the numbers in the data set
2. Divide the sum by the total numbers in the data set

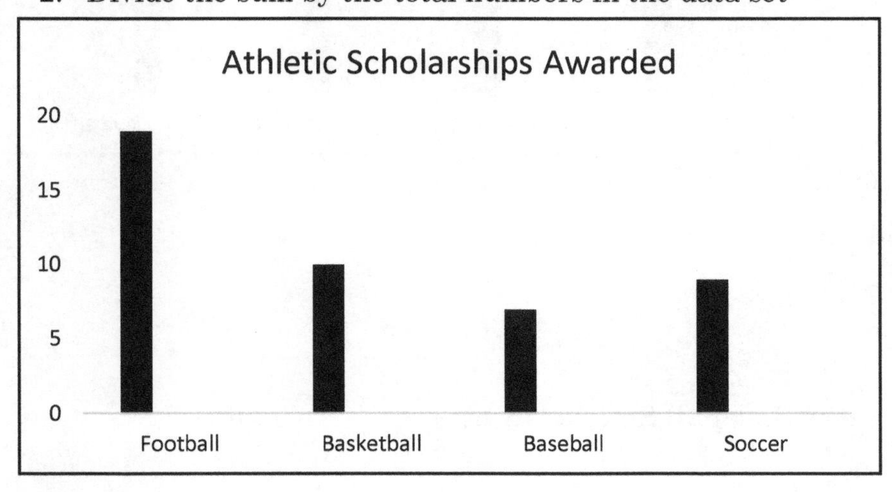

Values may not be stated explicitly stated in the graph. If so, you'll need to estimate as best you can.

Football: 19

Basketball: 10

Baseball: 7

Soccer: 9

Average: $\frac{19+10+7+9}{4} = 11.25$

> Sometimes answers may or may not include a decimal, depending on the context. If decimals are not given, then you should round to the nearest whole number

Example 2

Based on the graph, what percentage of the scholarships awarded were given to baseball players?

Step 1: Identify what the question is asking

Based on the graph, what **percentage of the scholarships** awarded were given to **baseball players?**

Step 2: Use the proper subskills to execute

Percentage: $\frac{\text{number of baseball scholarships}}{\text{total number of scholarships}}$

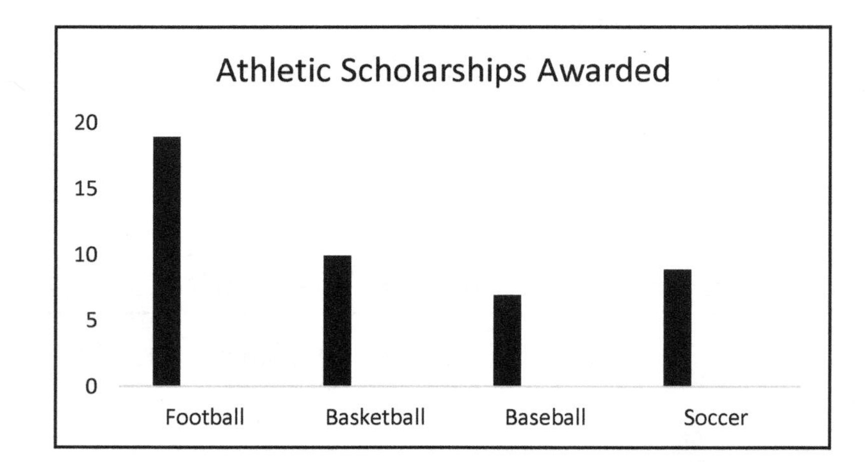

Football: 19

Basketball: 10

Baseball: 7

Soccer: 9

Percentage of Baseball Scholarships: $\frac{7}{19+10+7+9} = 15.6\%$

Circle Graph/Pie Chart

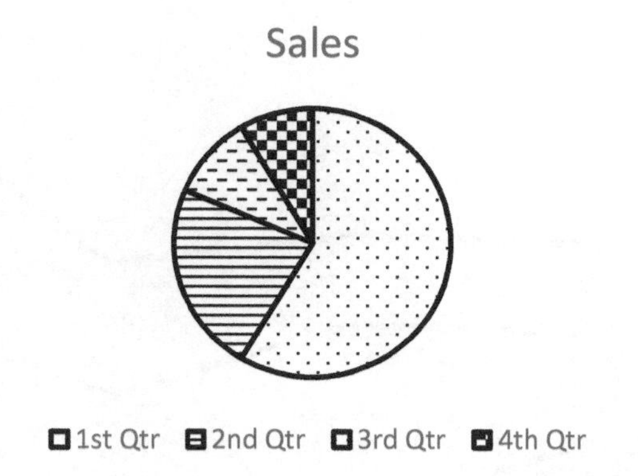

A circle graph is used to compare parts of data to the whole

When to use

- Generally, to show percentages or proportional relationships

Example 3

If there were 200 students surveyed, how many chose math as their favorite subject?

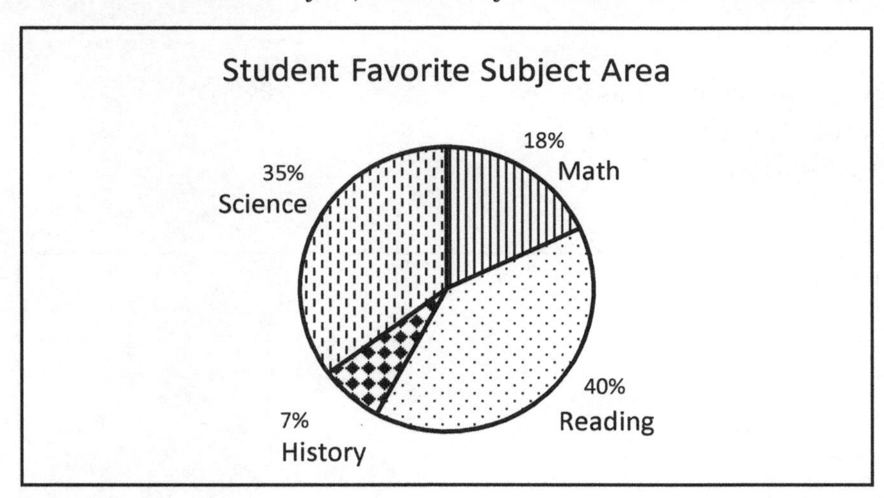

Step 1: Identify what the question is asking

If there were 200 students surveyed, **how many chose math** as their favorite subject?

Step 2: Use the proper subskills to execute

18% of students chose math

200 students total

18% of 200

$.18 \times 200 =$ **36 students**

Line Graph

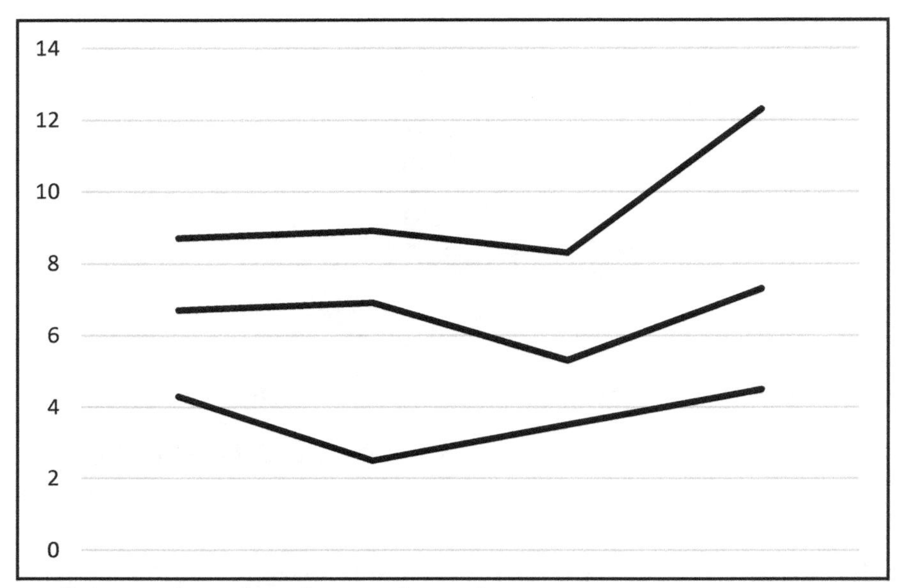

- A line graph is used to show how values change
  - The most common type of data found on a line graph illustrates how some variable **change over time**

When to use

- When you want to show trends
- When you want to make predictions
- When data is continuous

Scatter Plot

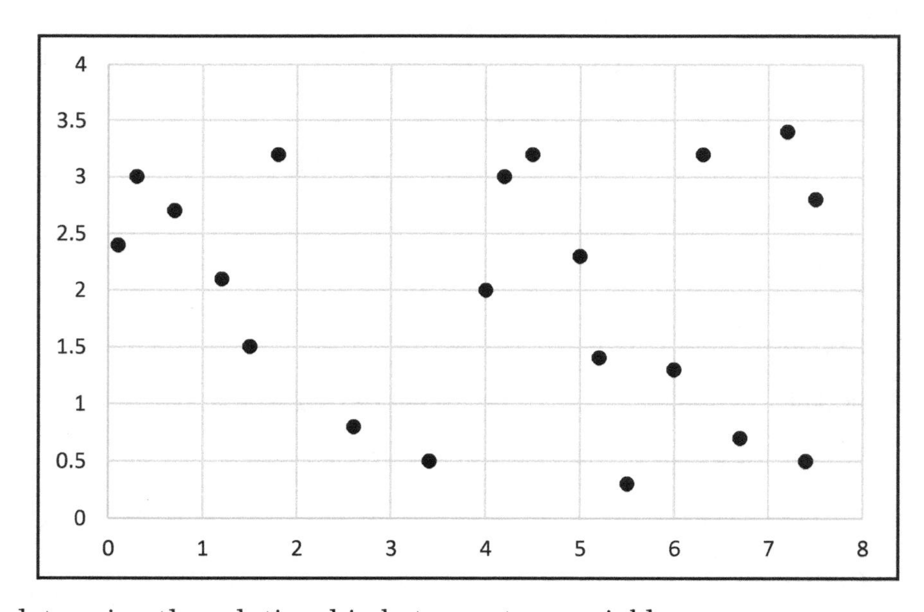

- Used to determine the relationship between two variables.
  - Variables can have a **positive** relationship, **negative** relationship or no relationship (pictured in graph above).

Review

1. What percentage of the varsity cheerleading squad is comprised of juniors?

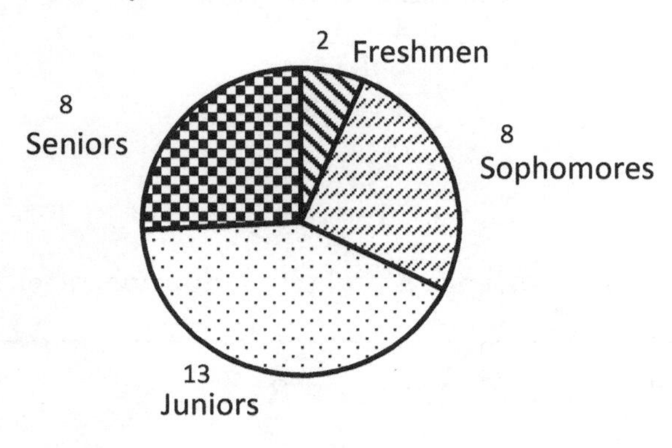

a. 42% b. 13% c. 26% d. 6%

2. Estimate the average number of Juniors and Seniors who made honor roll

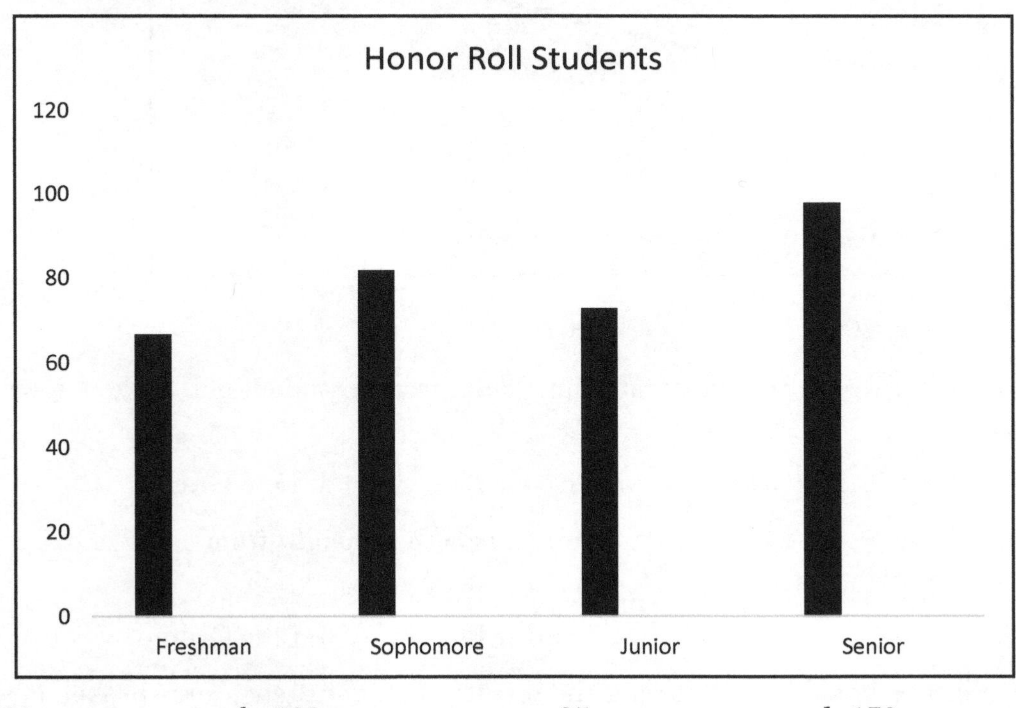

a. 300 b. 100 c. 85 d. 170

3. Which type of graph would best illustrate the relationship between the number of hours spent studying and test scores?

a. Bar Graph b. Pie Chart c. Scatter Plot d. Line Graph

4. What percentage ATM Entertainment's revenue do concerts account for?

ATM ENTERTAINMENT Revenue by event 2018	
Rap Shows	**$1,292,834**
R&B Shows	**$934,892**
Community Events	**$204,892**
Comedy Shows	**$452,987**
Sports Events	**$100,632**

a. 70% b. 35% c. 75% d. 30%

5. During which year did Florida experience a higher graduation rate than Texas?

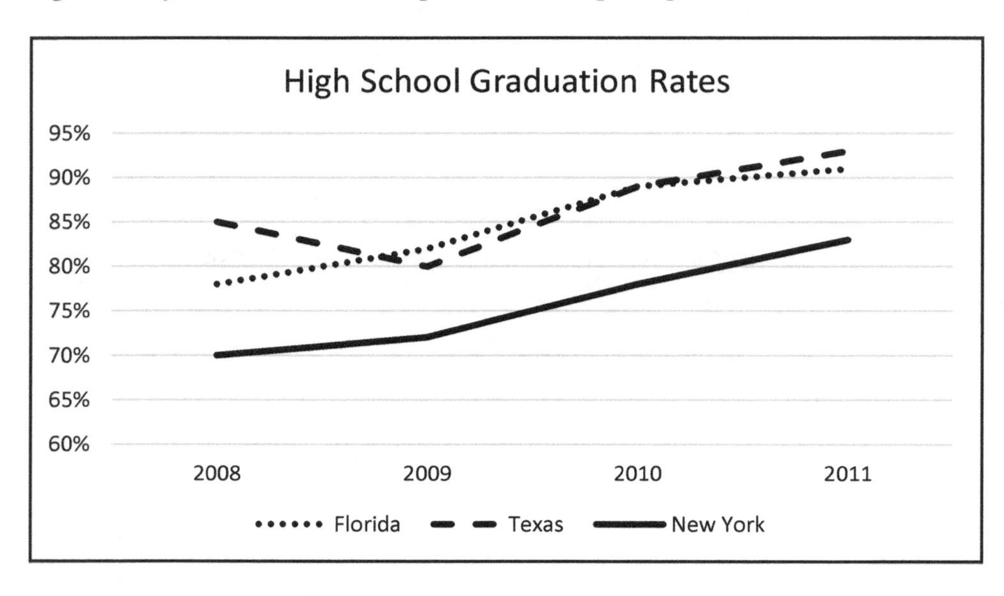

a. 2008 b. 2009 c. 2010 d. 2011

6. Which type of graph would best illustrate the results from a student poll broken down into percentages?

a. Bar Graph b. Pie Chart c. Scatter Plot d. Line Graph

7. Which type of graph would be illustrate a child's growth in height from age five to age fifteen?

a. Bar Graph b. Pie Chart c. Scatter Plot d. Line Graph

8. Which type of graph would best illustrate the results of student test scores broken down into letter grades?

a. Bar Graph b. Pie Chart c. Scatter Plot d. Line Graph

9. Which school experienced the largest decrease in student enrollment between the second and third semester?

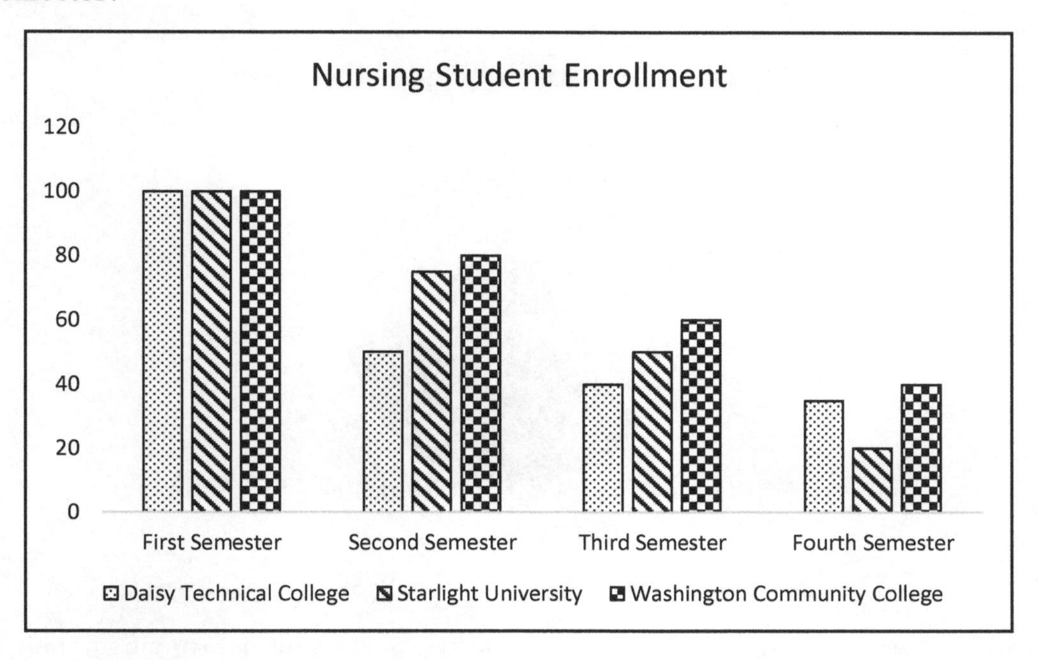

a. Daisy Technical College b. Starlight University c. Washington Community College

10. Which graph type best illustrates the relationship between two variables?

a. Bar Graph b. Pie Chart c. Scatter Plot d. Line Graph

11. What was the average number of students enrolled at Exceptional Studies Tutoring during the month of May?

Exceptional Studies Tutoring Student Enrollment by Subject Area May 2013	
Math	**58**
Reading	**43**
Science	**29**
Standardized Test Prep	**45**
Study Skills	**15**

a. 190 b. 58 c. 38 d. 40

12. Which type of graph would best illustrate data that a researcher describes as skewed to the right?

a. Bar Graph b. Histogram c. Scatter Plot d. Line Graph

13. Which type of graph would best illustrate student aptitude scores from the last ten years?

a. Bar Graph b. Pie Chart c. Scatter Plot d. Line Graph

14. By how much did House of JYL's sales increase in the 4th quarter?

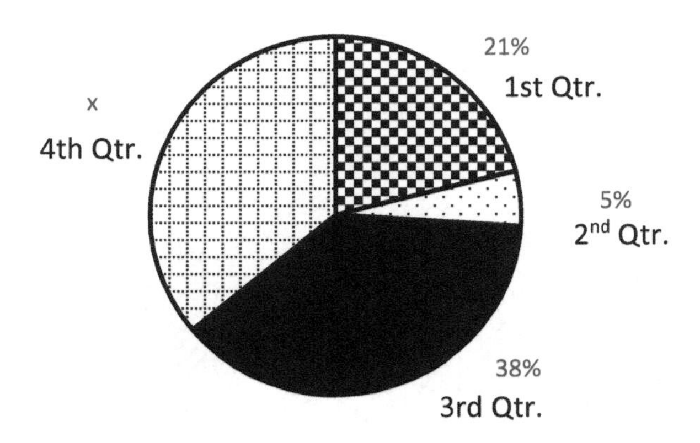

a. 36% b. 38% c. 21% d. 5%

15. If there were 500 people surveyed, how many chose sweet potato pie as their favorite dessert?

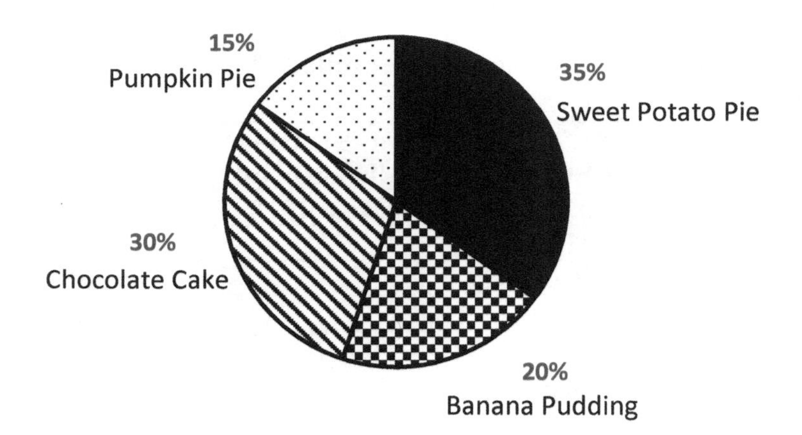

a. 35 b. 15 c. 175 d. 75

16. Which graph best illustrates the different types of costumes 50 students wore for Halloween?

a. Bar Graph b. Pie Chart c. Scatter Plot d. Line Graph

17. Which state exhibited the largest percentage growth in football recruits between 2016 and 2018?

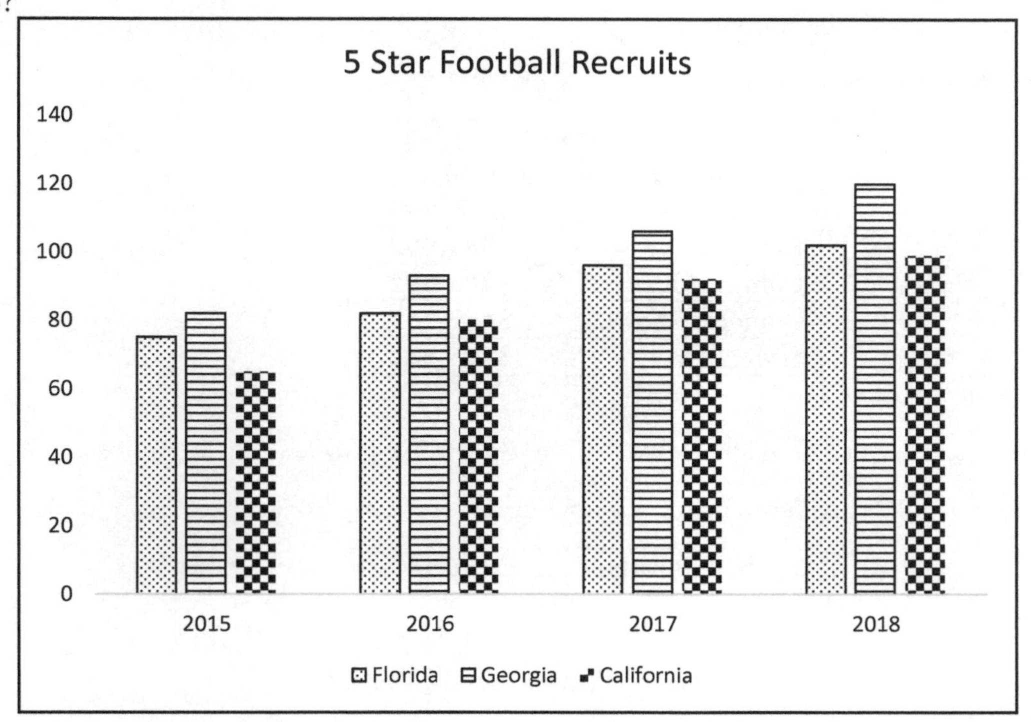

a. Florida b. Georgia c. California d. New York

18. During which year did two states have equal graduation rates?

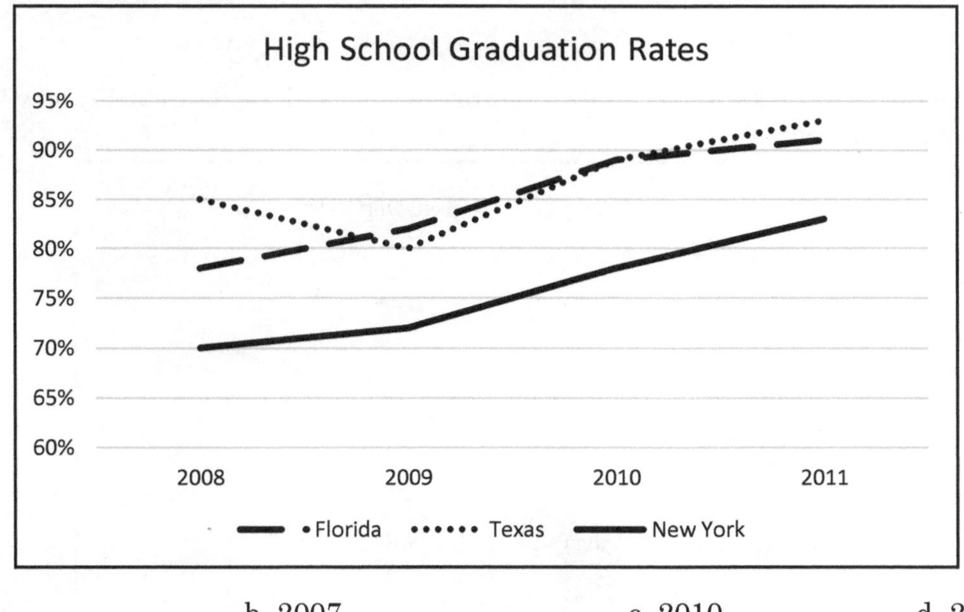

a. 2008 b. 2007 c. 2010 d. 2011

19. Which graph would best illustrate the relationship between the number of hours playing video games and standardized test scores?

a. Bar Graph b. Pie Chart c. Scatter Plot d. Line Graph

20. Which graph would best illustrate data that a researcher would describe as uniform?

a. Bar Graph b. Histogram c. Scatter Plot d. Line Graph

21. Which event accounts for roughly 7% of ATM Entertainment's revenue?

ATM ENTERTAINMENT Revenue by event 2018	
Rap Shows	**$1,292,834**
R&B Shows	**$934,892**
Community Events	**$204,892**
Comedy Shows	**$452,987**
Sports Events	**$100,632**

a. Community Events b. Rap Shows c. Sports Events d. Comedy Shows

22. If there are 200 students enrolled, how many students participate in football?

Student Sports Enrollment 2018	
Football	**52%**
Basketball	**34%**
Baseball	**6%**
Volley Ball	**3%**
Golf	**5%**

a. 52 b. 100 c. 104 d. 105

23. If there were 2,980,000 high school students enrolled in Florida in 2011, how many students did not graduate?

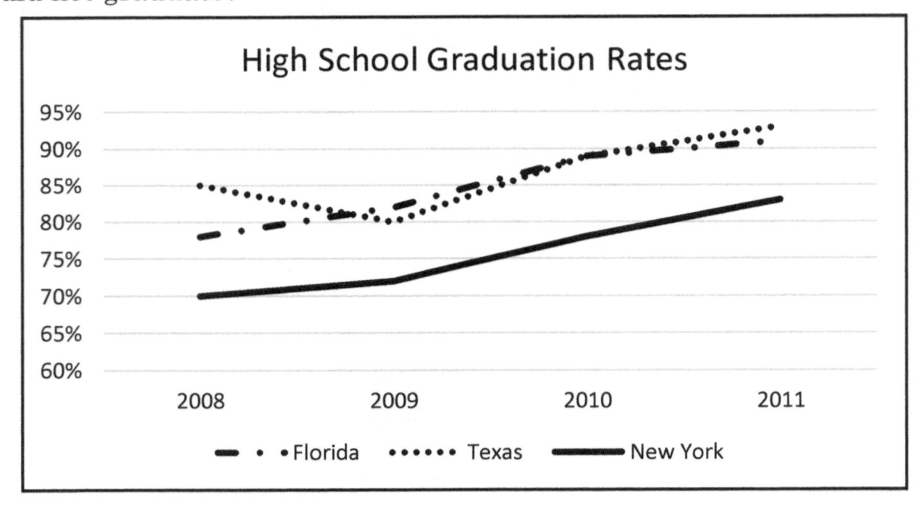

a. 2,682,000 b. 2,980,000 c. 298,000 d. 300,000

24. Based on the graph below, what percentage of the men surveyed chose red as their favorite car color?

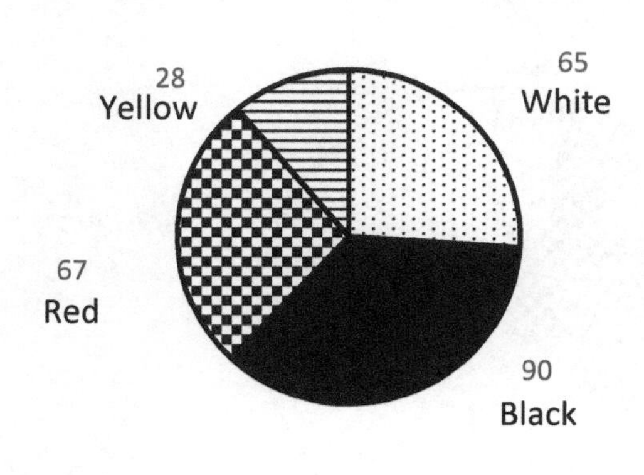

a. 67% b. 27% c. 90% d. 28%

25. Estimate the number of students who qualified for honor roll.

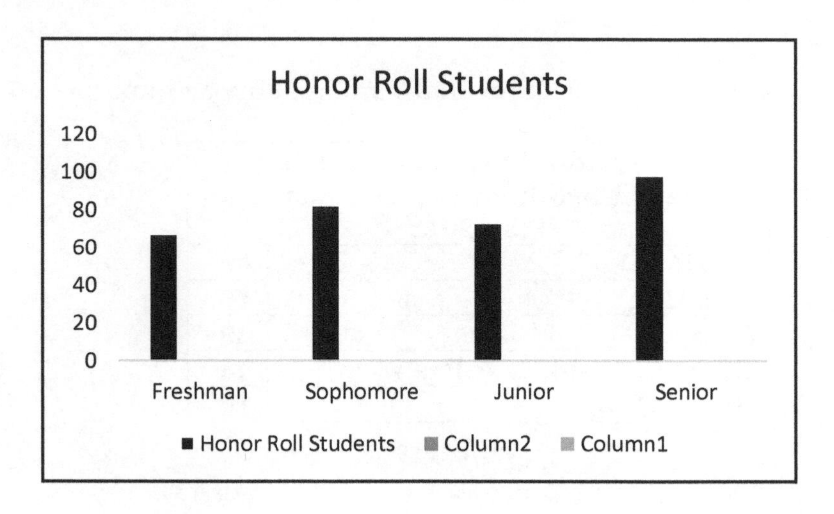

a. 300 b. 350 c. 250 d. 400

26. Which graph best illustrates the change in a variable over time?

a. Bar Graph b. Pie Chart c. Scatter Plot d. Line Graph

27. Which graph best illustrates the shape of a distribution?

a. Histogram b. Pie Chart c. Scatter Plot d. Line Graph

28. Which graph best illustrates data broken down into percentages?

a. Bar Graph b. Pie Chart c. Scatter Plot d. Line Graph

29. If House of JYL sold 1,000 items during the month of June, how many Jumpsuits were sold?

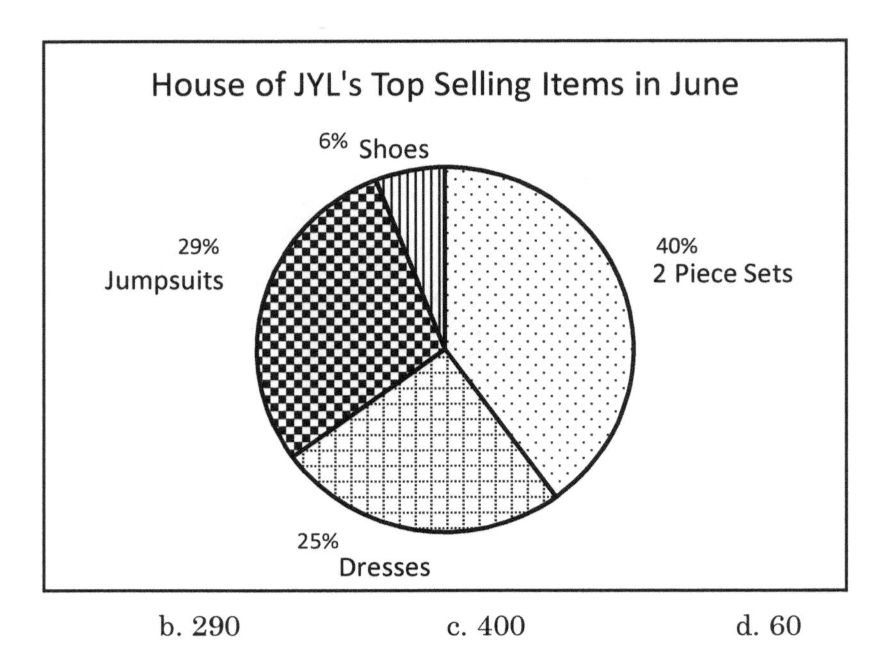

a. 250 b. 290 c. 400 d. 60

30. What percentage of student enrollment do math and reading account for?

Exceptional Studies Tutoring Student Enrollment by Subject Area May 2013	
Math	58
Reading	43
Science	29
Standardized Test Prep	45
Study Skills	15

a. 101% b. 58% c. 53% d. 43%

31. Which graph would best illustrate the results from a survey given to 1,000 students seeking information about their time spent on social media?

a. Bar Graph b. Pie Chart c. Scatter Plot d. Line Graph

32. Which graph would best illustrate a normal distribution?

a. Histogram b. Pie Chart c. Scatter Plot d. Line Graph

33. Which graph would best illustrate a state's graduation rates from the last five years?

a. Histogram b. Pie Chart c. Scatter Plot d. Line Graph

34. Which graph would best illustrate data that is considered to be discrete?

a. Bar Graph b. Pie Chart c. Scatter Plot d. Line Graph

35. Calculate the percentage decrease in student enrollment from the first to fourth semester at Starlight University?

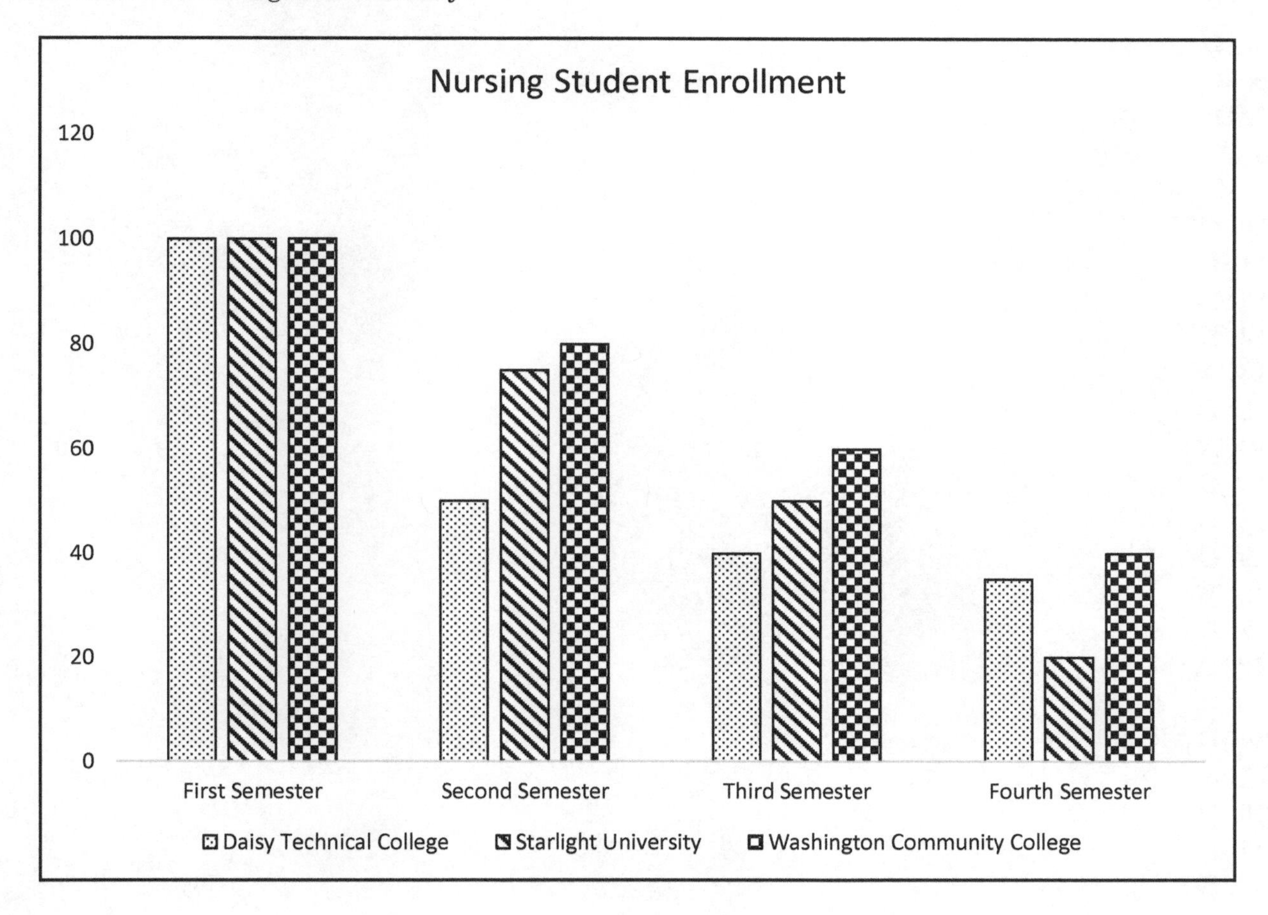

a. 80% b. 90% c. 50% d. 60%

Answer Key

1. A
2. C
3. C
4. C
5. B
6. B
7. D
8. A
9. B
10. C
11. C
12. B
13. D
14. A
15. C
16. A
17. B
18. C
19. C
20. B
21. A
22. C
23. C
24. B
25. A
26. D
27. A
28. B
29. B
30. C
31. A
32. A
33. D
34. A
35. A

Calculate Geometric Quantities

You will learn:

How to calculate the perimeter of a figure

How to calculate the area of a figure

How to solve for the missing dimension of a figure

Study Tips

Read and study EVERY example problem.

Complete EVERY practice problem.

Check to make sure all answers are correct.

Go back to correct the questions you answered incorrectly.

If you don't receive at least an 80% on the review, go back and practice the topic.

Calculating Geometric Quantities

Shapes

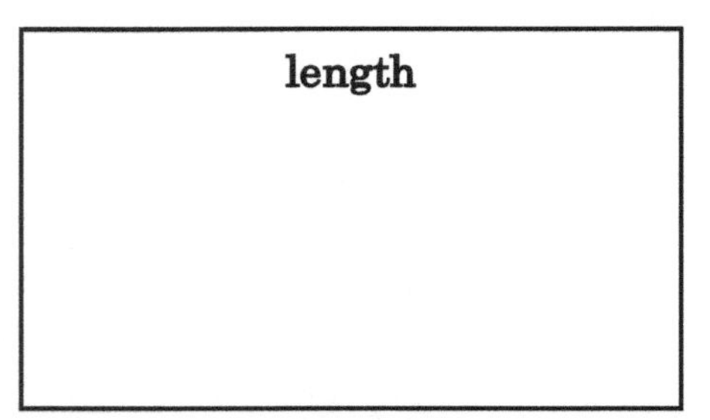

width

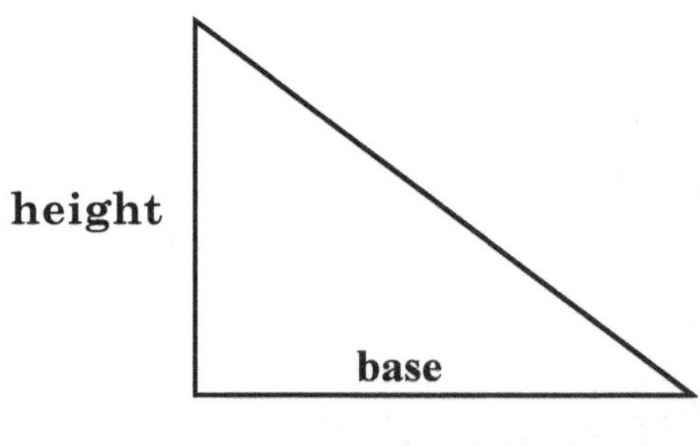

Area = length x width

Perimeter = the sum of all the sides

Area = $\frac{\text{base x height}}{2}$

Perimeter = the sum of all the sides

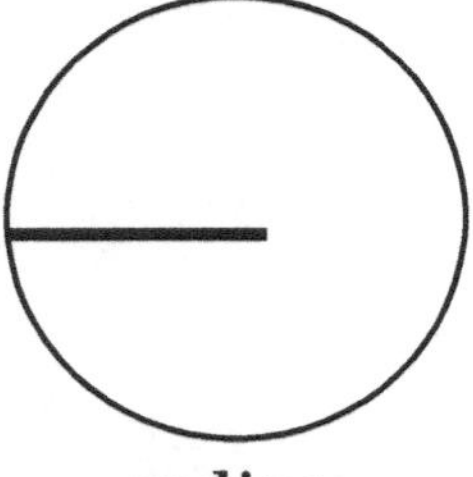

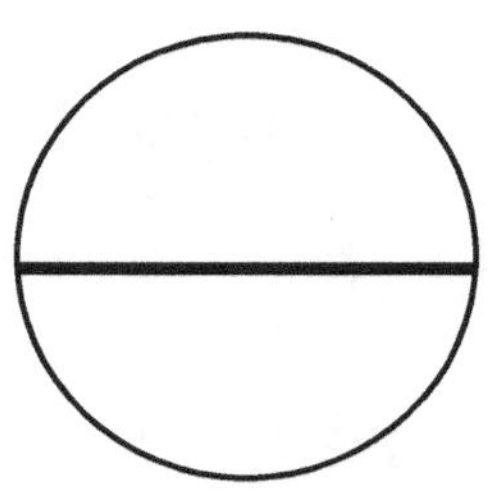

radius

diameter

Area = π x radius^2

Circumference = π x diameter

Circumference = 2 x π x radius OR 6.28 x π

(radius = $\frac{1}{2}$ diameter)

(π = 3.14)

side

Area = side × side

Perimeter = 4s

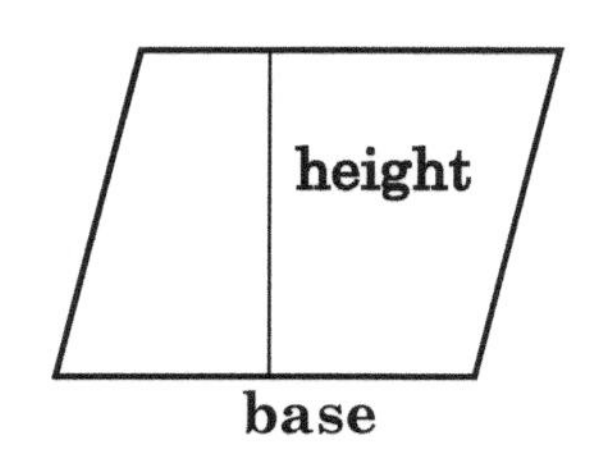

base

Area = base x height

Perimeter = the sum of all the sides

Calculating Perimeter

KISS IT

Step 1: Identify the given shape

Step 2: Apply the correct formula

Example 1

Lisa is choosing new tiles for her kitchen floor. If the tiles measure 12 inches on each side, what is the total distance around each tile?

Step 1: Identify given shape

Lisa is choosing new tiles for her kitchen floor. If the tiles measure **12 inches on each side**, what is the total distance around each tile?

↓ Squares have equal sides

Step 2: Apply the correct formula

Lisa is choosing new tiles for her kitchen floor. If the tiles measure 12 inches on each side, what is the total **distance around** each tile?

↓ **Perimeter**

Perimeter = side + side + side + side OR 4 × side

Perimeter = 12 + 12 + 12 + 12

Perimeter = 48 inches

Example 2

The bottom of a cup has a radius of 2 inches. What is the circumference of the cup's bottom in terms of pi?

Step 1: Identify given shape

What is the **circumference** of the cup's bottom in terms of pi?

Circle

***In terms of pi:** Do not convert pi to 3.14, leave it as is.

Step 2: Apply the correct formula

Circumference = $2 \times \pi \times r$

$= 2 \times 2 \times \pi$ **= 4π inches**

Practice 1

1. Anna's favorite box of cereal is larger than the average box with dimensions of 15 inches by 10 inches. What is the perimeter of the box's face?

2. If Kevin's bottle top has a radius of .75 cm, what is its circumference?

3. Alanah's favorite picture of her dad is 5 inches by 7 inches, what is the distance around the picture?

4. The top of Adrianne's jewelry box measures 14 centimeters on each side. What is the perimeter of the box's top?

5. Calculate the perimeter of the bedroom in the diagram below.

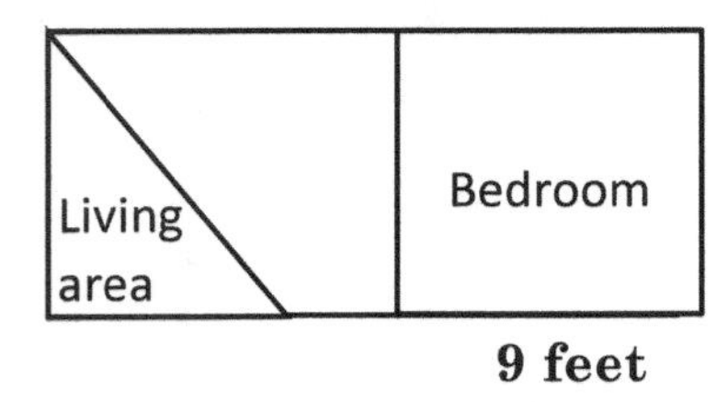

6. A slice of pizza measures 7 inches on two sides, and 6 inches on the third side. What is the perimeter of one slice?

7. Calculate the perimeter of the children's play area in the diagram below.

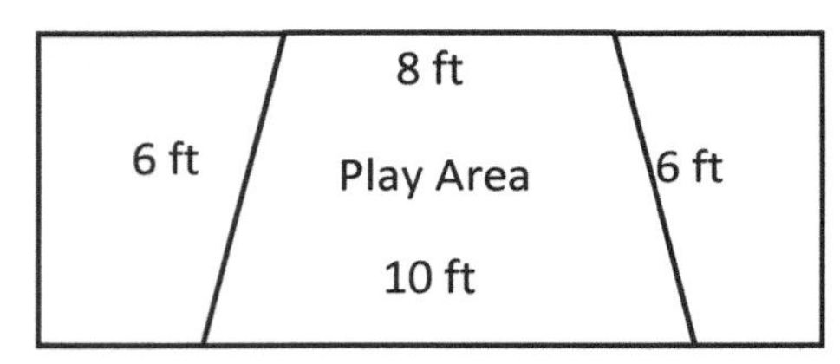

8. The bottom of a slice of square cheese measures $3\frac{1}{2}$ inches, what is its perimeter?

9. Nick's pizza place is home to the city's largest pizza pie with a diameter of 4 inches. What is the pie's circumference in inches?

10. The face of a television remote is 10 inches long and 2 inches wide. What is the distance around the remote's face?

Calculating Area

KISS IT

Step 1: Identify the given shape

Step 2: Apply the correct formula

Example 3

If a pizza has a radius of 10 inches, what is its area?

Step 1: Identify given shape

If a **pizza** has a radius of 10 inches, what is its area?

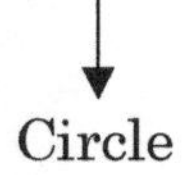

Circle

Step 2: Apply the correct formula

Area = pi × (radius)^2

Area = 3.14×10^2

$= 314$ in^2

Example 4

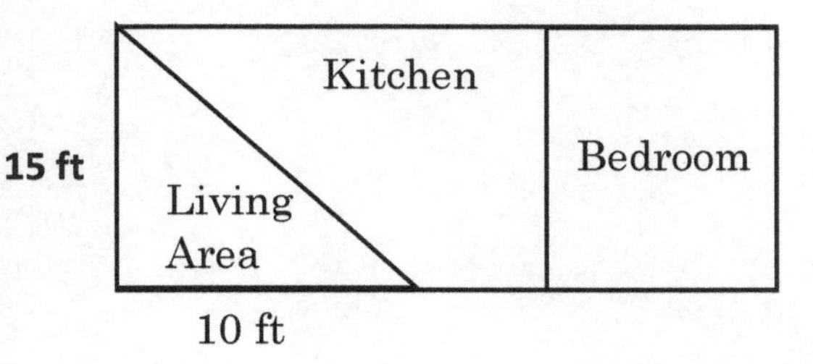

Based on the diagram above, how many square feet does the living area cover?

Step 1: Identify given shape

Based on the diagram above, how many square feet does the **living area** cover?

Right Triangle

Step 2: Apply the correct formula

Area = $\frac{\text{base x height}}{2} = \frac{10 \times 15}{2} = 75$ square feet

Practice 2

1. Anna's favorite box of cereal is larger than the average box with dimensions of 15 inches by 10 inches. What is the area of the box's face?

2. If Kevin's bottle top has a radius of .75 cm, what is its area? (round to nearest tenth)

3. Alanah's favorite picture of her dad is 5 inches by 7 inches, what is its area?

4. The top of Adrianne's jewelry box measures 14 centimeters on each side. What is the area of the box's top?

5. What is the area of the bedroom in the diagram below?

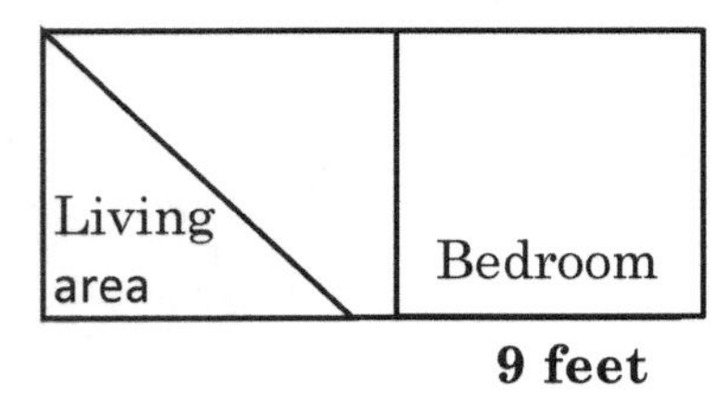

6. Calculate the area of the figure below.

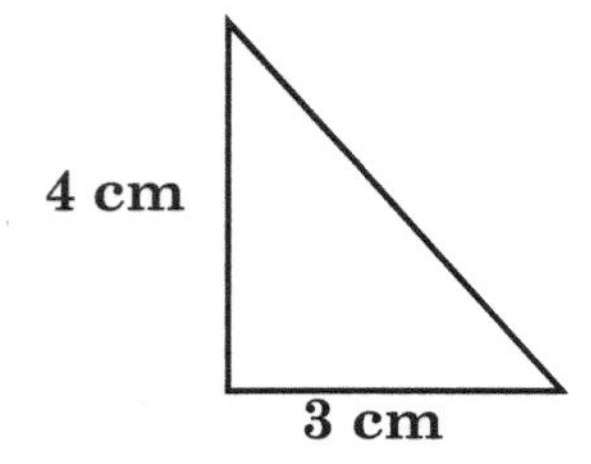

7. Calculate the area of the diagram below

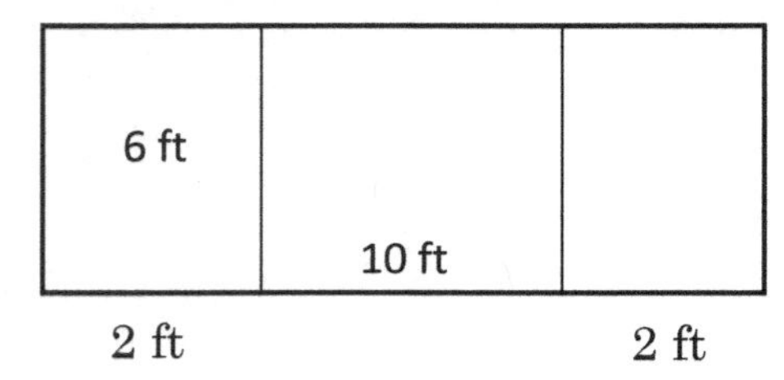

8. The bottom of a slice of square cheese measures $3\frac{1}{2}$ inches, what is its area?

9. Nick's pizza place is home to the city's largest pizza pie with a diameter of 4 feet. What is the pie's area in inches?

10. The face of a television remote is 10 inches long and 2 inches wide. What is the area of the remote's face?

Solve for Missing Dimensions

KISS IT

Step 1: Identify the given shape

Step 2: Apply the correct formula

Example 5

The area of the basketball court at the local gymnasium is 1200 square feet. If the length of the court is 40 feet, what is the width?

Step 1: Identify given shape

The area of the **basketball court** at the local gymnasium is 1200 square feet.

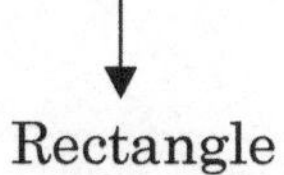

Rectangle

Step 2: Apply the correct formula

Area = length × width

1200 = 40 × width

30 inches = width

Example 6

If the circumference of a table top is 75 inches, what is its diameter (to the nearest tenth)?

Step 1: Identify given shape

If the **circumference** of a table top is 75 inches, what is its diameter?

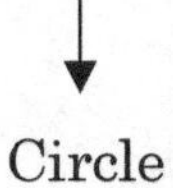

Circle

Step 2: Apply the correct formula

Circumference = π × d

75 = 3.14 × d

23.9 inches = d

Practice 3

1. Anna's favorite box of cereal is larger than the average box with a length of 17 inches. If the area of the box's face is 153 square inches, what is its width?

2. If Kevin's bottle top has an area of 2.25 cm^2, what is its radius? (round to the nearest hundredth)

3. Alanah's favorite picture of her dad covers 84 square inches of space on the wall. If the length of the picture is 12 inches, what is its width?

4. The top of Adrianne's jewelry box has an area of 81 square centimeters. If the box's top is shaped like a square, what is the measure of each side?

5. Calculate the dimensions of the bedroom in the diagram below.

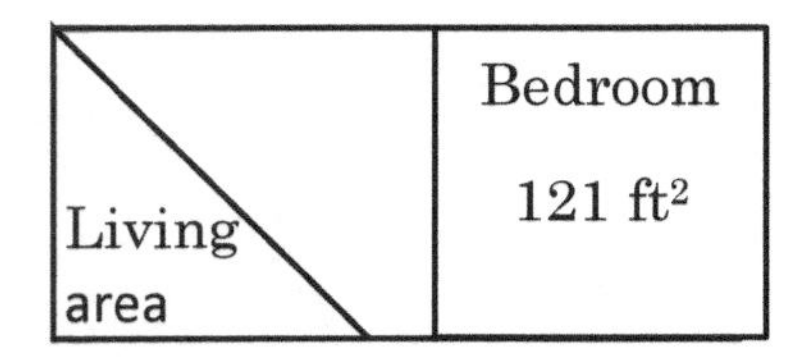

6. Calculate the missing dimension in the figure below

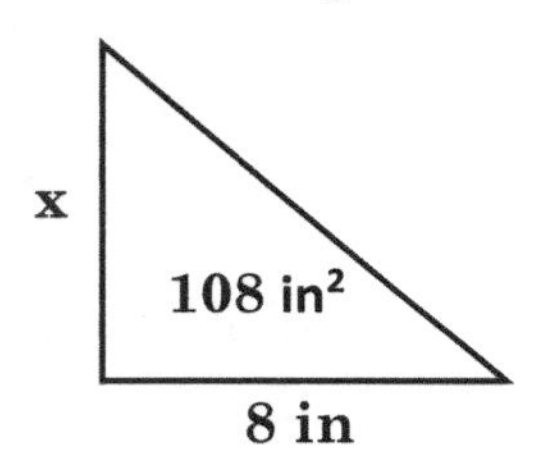

7. A circle has a circumference of 100 inches, what is its radius? (round to the nearest tenth)

8. Calculate the missing dimension in the figure below (nearest hundredth)

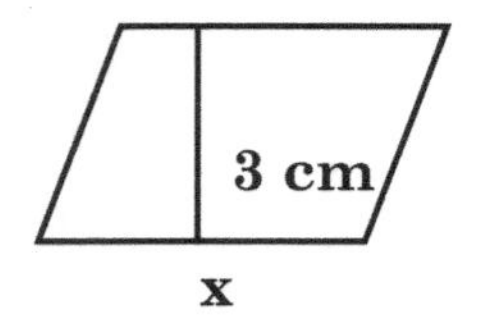

Area: 32 cm^2

9. Nick's pizza place is home to the city's largest pizza pie with an area of 85 square inches. What is the pie's radius? (round to the nearest hundredth)

10. The face of a television remote has an area of 24 square inches. Calculate the length of the remote if the width measures 1.5 inches.

Review

1. Karla's dorm room measures 10 feet by 12 feet. Calculate the distance around her room.

2. Taylor needs to measure the distance around her television, but she doesn't have a tape measure. She knows that the television is 42 inches long and 36 inches tall. Calculate the distance around the television.

3. Brittany's custom denim jacket is filled with buttons from the various countries she's visited. She has 50 square centimeters of space left for one button. If her last button has a radius of 5 cm, will it fit in the available space?

4. Calculate the area and perimeter of the figure below

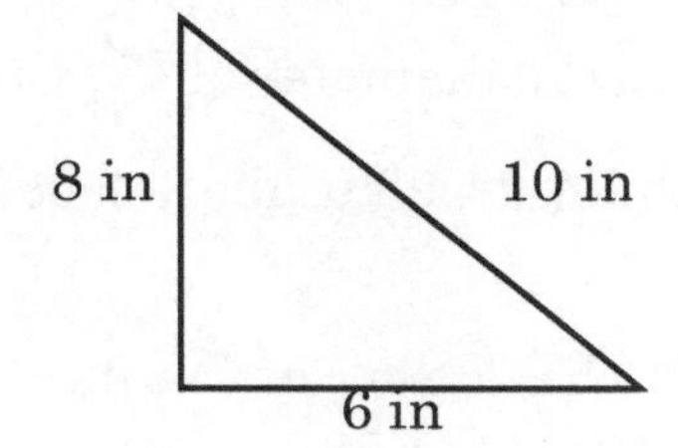

5. If a square-shaped tile covers 12.25 square feet, what is the measure of each side?

6. The radius of a circle measures 1.5 inches. Calculate the circumference and area of the figure.

7. Karla's dorm room measures 10 feet by 12 feet. Calculate the square footage of her room.

8. Calculate the area and circumference of the figure below

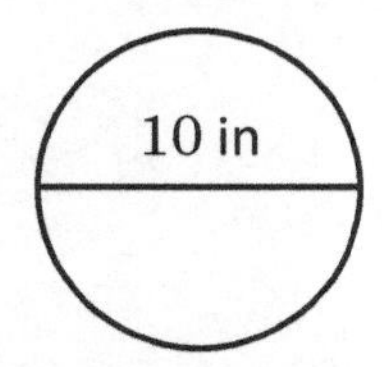

9. If the area of a swimming pool measures 1,000 square feet and the length has been measured as 50 feet, what is the width?

10. Taylor needs to measure the amount of space her television covers on the wall, but she doesn't have a tape measure. She knows that the television is 42 inches long and 36 inches tall. Calculate the amount of space the television covers.

11. The radius of a circle measures 5 inches. What is the area and circumference in terms of pi?

12. Melanie's spare tire measures 15 inches across the center. Calculate the area of the tire?

13. A standard door measures $6\frac{1}{2}$ feet tall and $3\frac{1}{2}$ feet wide. Calculate the area and perimeter.

14. Calculate the height of the figure below

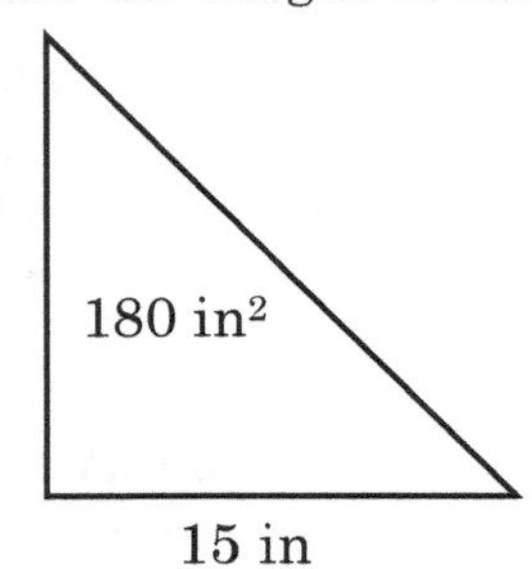

15. If Karla's dorm room has a square footage of 200 feet and her longest wall measures 15 feet, what is the length of the shortest wall? (Assume the room is rectangular in shape).

16. Melanie's spare tire has an area of 200 inches^2. Calculate the diameter of the tire.

17. A windowpane measures 6 feet by 4 feet. If windows account for 288 square feet in Lisa's house, how many windows are there?

18. Taylor needs to calculate the length of her television, but she doesn't have a tape measure. She knows that the television is 42 inches long and covers 1,596 square inches. Calculate the width of the television.

19. Melanie's spare tire measures 15 inches across the center. Calculate the circumference of the tire in terms of pi?

20. Solve for the missing dimensions of the figure below

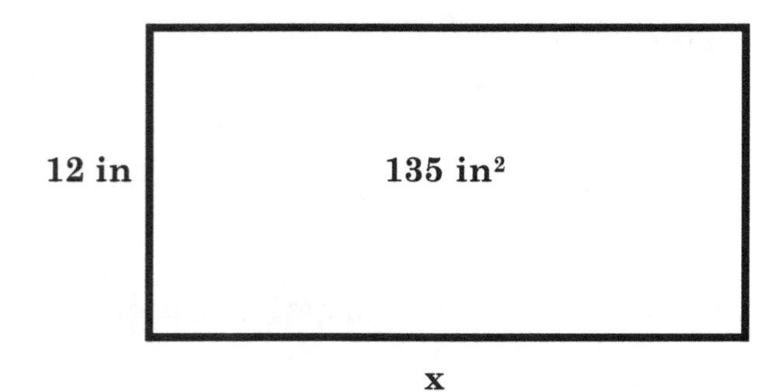

Answer Key

Practice 1

1. 50 in
2. 4.71 cm
3. 24 in
4. 56 cm
5. 36 ft
6. 20 in
7. 30 ft
8. 14 in
9. 12.56 in
10. 24 in

Practice 2

1. 150 in^2
2. 1.8 cm^2
3. 35 in^2
4. 196 cm^2
5. 81 ft^2
6. 6 cm^2
7. 84 ft^2
8. 12.25 in^2
9. 12.56 in^2
10. 20 in^2

Practice 3

1. 9 in
2. .85 cm
3. 7 in
4. 9 cm
5. 11 ft
6. 27 in
7. 15.9 in
8. 10.67 cm
9. 5.20 in
10. 16 in

Review

1. 44 ft
2. 156 in
3. No
4. Area: 24 in^2 Perimeter: 24 inches
5. 3.5 ft
6. Circumference: 9.42 in Area: 7.07 in^2
7. 120 ft^2
8. Circumference: 31.4 in Area: 78.5 in^2
9. 20 ft
10. 1,512 in^2
11. Circumference: 10π Area: 25π
12. 176.63 in^2
13. Perimeter: 20 ft Area: 22.75 ft^2
14. 24 in
15. 13.33 ft
16. 15.96 in
17. 12 windows
18. 38 in
19. 15π
20. 11.25 in

BLANK PAGE

Cluster 3 Review

You should know:

☐ How to identify independent variables

☐ How to identify dependent variables

☐ How to identify correlation given a graph

☐ How to identify correlation given a scenario

☐ How to calculate mean

☐ How to calculate median

☐ How to calculate mode

☐ How to calculate range

☐ How to classify the shape of a distribution

☐ How to identify the graph that best illustrates a scenario

☐ How to calculate statistics given a graph

☐ How to calculate percentages given a graph

☐ How to reach valid conclusions given a graph

☐ How to calculate the perimeter of a figure

☐ How to calculate the area of a figure

☐ How to solve for the missing dimension of a figure

Cluster 3 Review

1. Identify the median of the given data set: 19 22 17 28 23

a. 21.8 b. 22 c. 11 d. 20

2. A circle has a radius of 5 cm. What is its area in terms of pi?

a. 5п b. 25п c. 10п d. 2.5п

3. As students study more hours, their test scores increase. Identify the dependent variable

a. students b. hours c. test scores d. study

4. The graph below represents college students' favorite things to do in their spare time. Based on the graph below, what percentage of students play video games?

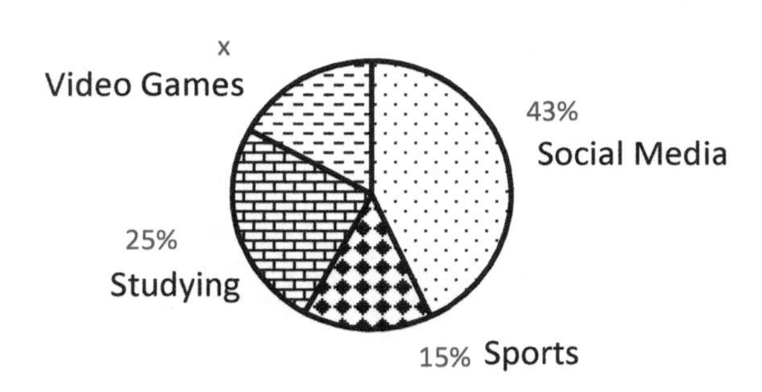

a. 43% b. 71% c. 30% d. 17%

5. Describe the relationship between number of hours worked and salary

a. no correlation b. negative correlation c. positive correlation

6. Calculate the mean of the given data set: 22 25 13 19 20 22

a. 20.2 b. 21 c. 12 d. 22

7. Calculate the base of the figure below

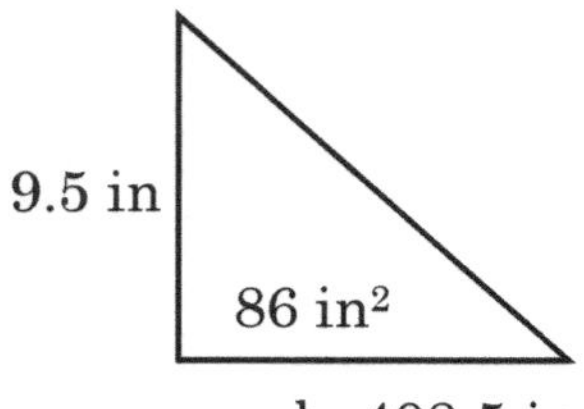

a. 817 in b. 408.5 in c. 18 in d. 9 in

8. As athletes age, their bones weaken. Identify the independent variable

a. athletes b. bones c. weaken d. age

9. The top of Alice's shoe box measures 12 inches by 7 inches. If she wants to cover the top in decorative paper, how much will she need?

a. 38 in b. 84 in^2 c. 38 in^2 d. 84 in

10. Which type of graph best illustrates change over a period of time?

a. circle graph b. histogram c. line graph d. bar graph

11. Based on the graph below, between which two years did House of JYL experience the most growth?

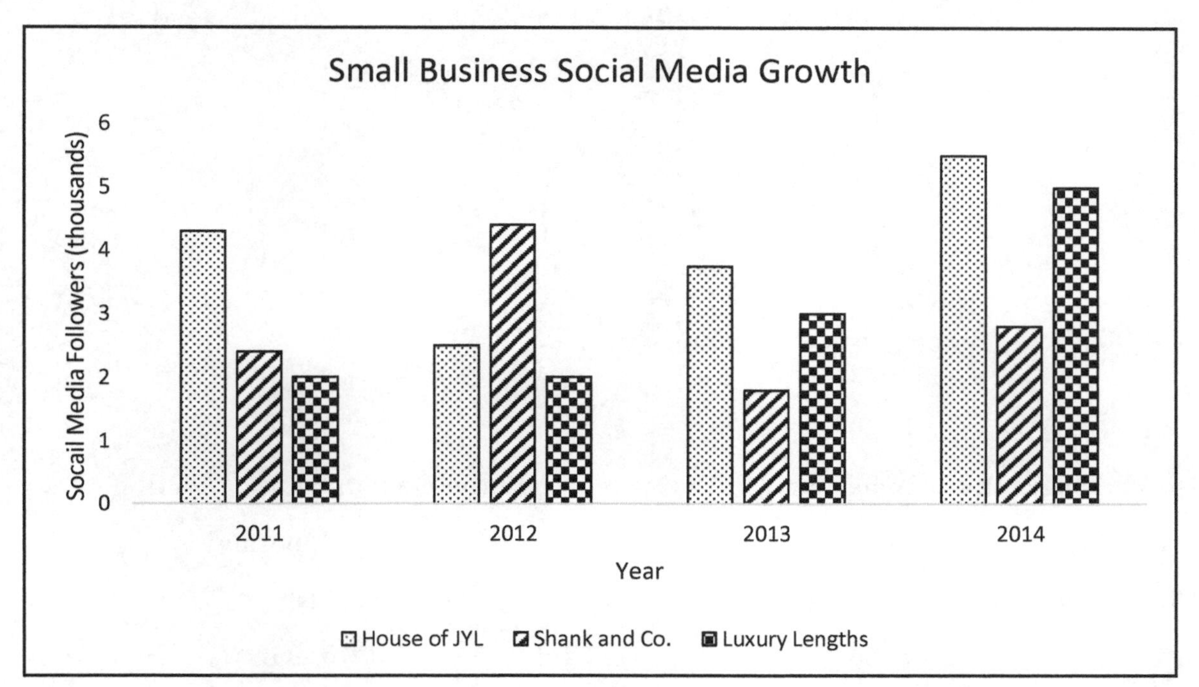

a. 2011 – 2012 b. 2012 – 2013 c. 2013 – 2014

12. Identify the mode of the given data set: 22 25 13 19 20 22

a. 22 b. 20.2 c. 21 d. 8

13. Calculate the circumference in terms of pi

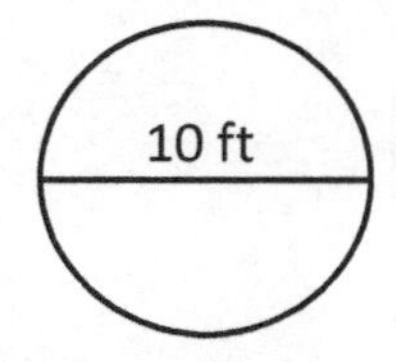

a. 5п b. 20п c. 25п d. 10п

14. Tourism decreases when gas prices rise. Identify the dependent variable

a. gas b. price c. decrease d. tourism

15. Which type of graph best illustrates the relationship between two variables?

a. line graph b. scatter plot c. circle graph d. line graph

16. Calculate the range of the given data set: 2 1 2 8 10 0

a. 10 b. 3.8 c. 2 d. 8

17. If there are 80 football players, how many of them are seniors?

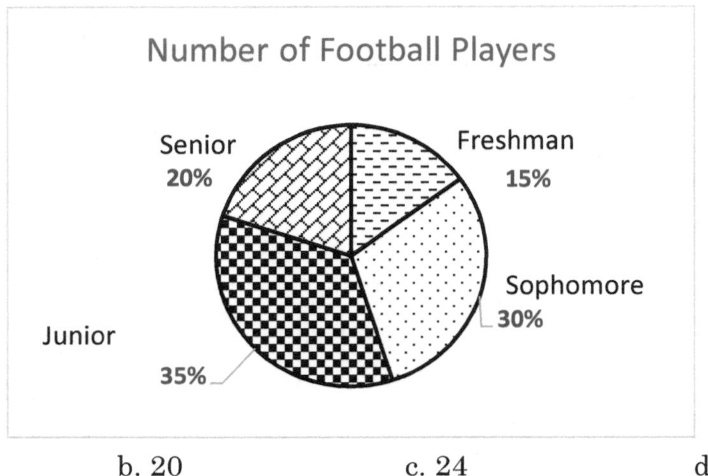

a. 16 b. 20 c. 24 d. 30

18. Which graph shape has a mean that is less than the median?

a. positive b. uniform c. negative d. normal

19. A square has an area of 49 in^2. What is the measure of one side?

a. 12.25 in b. 24.5 in c. 7 in d. 9 in

20. What type of relationship does this graph illustrate?

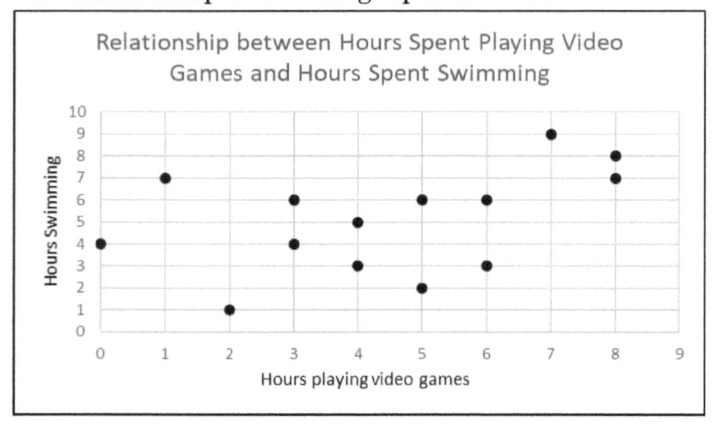

a. positive correlation b. no correlation c. negative correlation

Answer Key

1. B

2. B

3. C

4. D

5. A

6. A

7. C

8. D

9. B

10. C

11. C

12. A

13. D

14. D

15. B

16. A

17. A

18. C

19. C

20. B

Cluster 3 Review Break Down

Explain the Relationship Between Two Variables

Question 3 ☐

Question 5 ☐

Question 8 ☐

Question 14 ☐

Question 20 ☐

Evaluate Information in Tables Charts and Graphs Using Statistics

Question 1 ☐

Question 6 ☐

Question 12 ☐

Question 16 ☐

Question 18 ☐

Interpret Relevant Information from Tables Charts and Graphs

Question 4 ☐

Question 10 ☐

Question 11 ☐

Question 15 ☐

Question 17 ☐

Calculate Geometric Quantities

Question 2 ☐

Question 7 ☐

Question 9 ☐

Question 13 ☐

Question 19 ☐

Clusters 1 - 3

Apply Estimation Strategies and Rounding Rules to Real-World Problems

Convert Among Non-Negative Fractions, Decimals, and Percentages

Compare and Order Rational Numbers

Convert Within and Between Standard and Metric Systems

Solve Real-World Problems Involving Ratios and Rate of Change

Solve Real-World Problems Involving Proportions

Solve Real-World Problems Involving Percentages

Explain the Relationship Between Two Variables

Evaluate Information in Tables Charts and Graphs Using Statistics

Interpret Relevant Information from Tables Charts and Graphs

Calculate Geometric Quantities

Clusters 1-3 Review

You should know:

☐ How to round whole numbers and decimals

☐ How to simplify a radical

☐ How to choose an appropriate tool of measure

☐ How to estimate the mass, length, width, etc.... of everyday objects

☐ How to convert fractions to decimals

☐ How to convert fractions to percentages

☐ How to convert decimals to percentages

☐ How to convert decimals to fractions

☐ How to convert percentages to fractions

☐ How to convert percentages to decimals

☐ How to compare rational values

☐ How to identify the largest value in a data set

☐ How to identify the smallest value in a data set

☐ How arrange values in ascending order

☐ How to arrange values in descending order

☐ How to convert within standard and metric systems

☐ How to convert between standard and metric systems

☐ How to write the ratio between two quantities

☐ How to identify equivalent ratios

☐ How to solve for missing values given a ratio

☐ How to solve proportions with whole numbers

☐ How to solve proportions with fractions and decimals

☐ How to solve proportions containing values with different units

☐ How to calculate percentage increase

☐ How to calculate percentage decrease

☐ How to calculate tax, tip & discount

☐ How to calculate percent markups

☐ How to calculate percent markdowns

☐ How to solve percent proportions

☐ How to identify independent variables

☐ How to identify dependent variables

☐ How to identify correlation given a graph

☐ How to identify correlation given a scenario

☐ How to calculate mean

☐ How to calculate median

☐ How to calculate mode

☐ How to calculate range

☐ How to classify the shape of a distribution

☐ How to identify the graph that best illustrates a scenario

☐ How to calculate statistics given a graph

☐ How to calculate percentages given a graph

☐ How to reach valid conclusions given a graph

☐ How to calculate the perimeter of a figure

☐ How to calculate the area of a figure

☐ How to solve for the missing dimension of a figure

Cluster 1-3 Review

1. Which unit is best when estimating the length of a coffee table?

a. Feet b. inches c. ounces d. liters

2. Estimate $\sqrt{8}$

a. 2.83 b. 2.82 c. 2.5 d. 2.9

3. Estimate the length of a twin sized mattress

a. 75 cm b. 75 in c. 75 ft d. 75 m

4. Convert $\frac{6}{29}$ to a decimal

a. 4.84 b. .206 c. 4.83 d . 0.207

5. Convert 125% to a decimal

a. 125 b. 1.25 c. 12.5 d. 1250

6. Convert 0.85 to a fraction in simplest form

a. $\frac{17}{20}$ b. $\frac{85}{1000}$ c. $\frac{85}{10}$ d. $\frac{20}{17}$

7. Identify the smallest value: -12 -13 -14 -10

a. -12 b. -13 c. -14 d. -10

8. Arrange the values in descending order: 42% $\sqrt{.36}$ $\frac{3}{8}$ .376

a. $\sqrt{.36}$ 42% .376 $\frac{3}{8}$

b.42% .376 $\sqrt{.36}$ $\frac{3}{8}$

c. $\sqrt{.36}$ 42% $\frac{3}{8}$ .376

d. $\frac{3}{8}$ .376 42% $\sqrt{.36}$

9. Arrange the values in ascending order: $-\frac{1}{4}$ $\sqrt{9}$ -2 160%

a.$\sqrt{9}$ 160% $-\frac{1}{4}$ -2

b. -2 $-\frac{1}{4}$ 160% $\sqrt{9}$

c. $\sqrt{9}$ 160% -2 $-\frac{1}{4}$

d. $-\frac{1}{4}$ -2 160% $\sqrt{9}$

10. $2\frac{1}{4}$ yards = ____________ inches

a. 81 b. 6.75 c. 18 d. 20.25

11. .086 m = ________________mm

a. 8.6 b. .86 c. 860 d. 86

12. 6.89 miles = _____________km **(1 mile = 1.609 km)**

a. 4.28 b. 11.09 c. 4.29 d. 11.08

13. The ratio of men to women at Brandon's gym is 1:4. If the local gym has an equivalent ratio, identify a possible ratio from the selection below.

a. 4 to 7 b. 6 to 24 c. 24 to 6 d. 7 to 4

14. A bag includes 6 blue marbles, 4 red marbles, 7 yellow marbles, and 5 green marbles. What is the ratio of blue and red marbles to the total number of marbles in the bag?

a. $\frac{10}{21}$ b. $\frac{5}{11}$ c. $\frac{3}{2}$ d. $\frac{11}{5}$

15. The ratio of boys to girls in the school choir is 1 to 3. If there is a total of 21 girls in the choir, how many boys are there?

a. 21 b. 63 c. 7 d. 8

16. There are 60 football players, 35 basketball players, and 28 baseball players visiting the local college for a sports camp. What is the ratio of basketball players to football players?

a. 25:60 b. 7:12 c. 12:7 d. 60:35

17. Alicia can bake 10 cakes in one day. At this rate, how many cakes can she bake in 2 weeks?

a. 20 b. 70 c. 100 d. 140

18. Alexis can type $3\frac{1}{4}$ pages in 2 hours. At this rate, how many pages can she type in 6 hours?

a. $6\frac{1}{2}$ b. $9\frac{3}{4}$ c. $3\frac{9}{13}$ d. $18\frac{1}{4}$

19. If Stephanie's nephew can eat $1\frac{1}{2}$ pizzas in 2 hours, how many can he eat in 5 hours?

a. $\frac{3}{5}$ b. 3 c. $3\frac{3}{4}$ d. 4

20. Ashley purchased 8 pairs of shoes in two weeks. If she continues to shop at this rate, how many shoes will she purchase in 3 months?

a. 16 b. 9 c. 32 d. 48

21. Ashley's gross pay last year was $46,986. If she paid 15% in taxes, how much was her take home pay?

a. $39,938 b. $54,034 c. $7,048 d. $46,986

22. Employees at Shoe World receive a discount of 20% off very purchase. If one employee spends $75 and another employee spends $100, how much money will they receive off their purchases together?

a. $35 b. $15 c. $80 d. $5

23. Aaliyah always tips the waitress 15%. If her total bill came to $23.49, how much of a tip will she leave?

a. $27.01 b. $23.49 c. $3.52 d. $3.50

24. Gina purchased a pair of denim shorts that were on sale for $25. If the original price was $40, what was the percent decrease in price?

a. 37.5% b. 16% c. 40% d. 60%

25. As students study more hours, their test scores increase. Identify the dependent variable

a. students b. hours **c.** test scores d. study

26. What type of relationship does this graph illustrate?

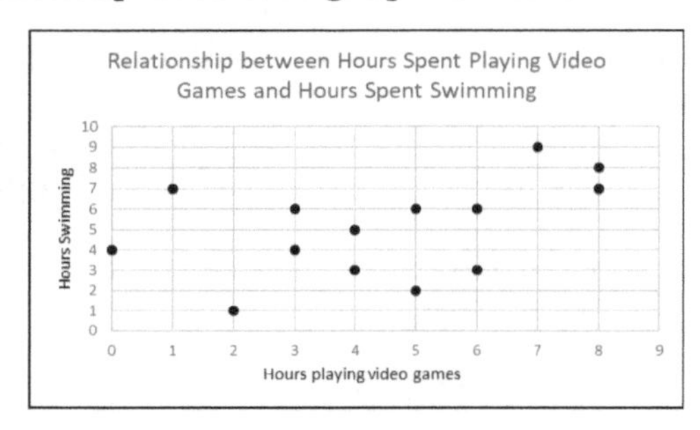

a. positive correlation b. no correlation c. negative correlation

27. As athletes age, their bones weaken. Identify the independent variable.

a. athletes b. bones c. weakens d. age

28. If there are 110 football players, how many of them are seniors?

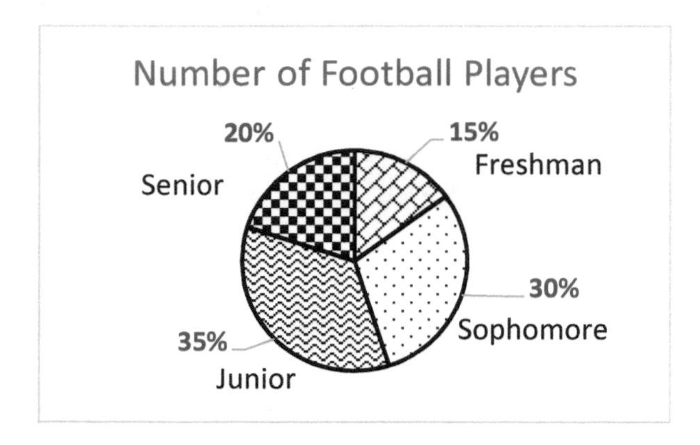

a. 16 b. 22 c. 24 d. 33

29. Which graph would best illustrate a normal distribution?

a. Histogram b. Pie Chart c. Scatter Plot d. Line Graph

30. What percentage of student enrollment do math and reading account for?

Exceptional Studies Tutoring Student Enrollment by Subject Area May 2013	
Math	**58**
Reading	**43**
Science	**29**
Standardized Test Prep	**45**
Study Skills	**15**

a. 101% b. 58% c. 53% d. 43%

31. Which graph shape has a mean that is greater than the median?

a. positive b. negative c. uniform d. normal

32. Calculate the mean of the given data set: 24 10 35 39 19

a. 24 b. 25.4 c. 29 d. 19

33. Calculate the median of the given data set: 24 10 35 39 19

a. 24 b. 25.4 c. 29 d. 19

34. A square has an area of 225 in². What is the measure of one side?

a. 12.25 in b. 56.25 in c. 15 in d. 112.5 in

35. A circle has a diameter of 10 cm. What is its area in terms of pi?

a. 5π b. 25π c. 10π d. 2.5π

36. Solve for the missing dimensions of the figure below

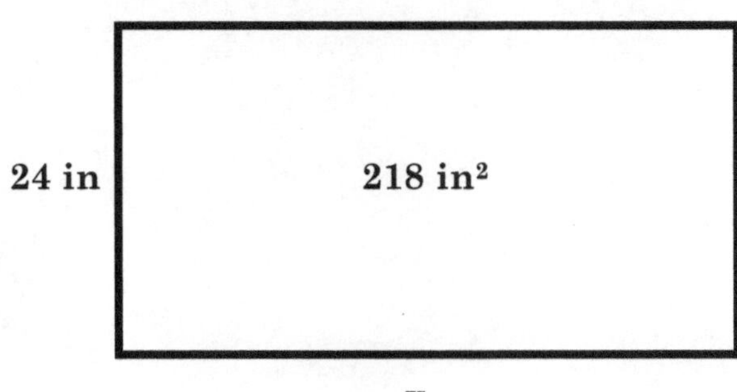

a. 54.5 b. 122 c. 9.08 d. 10

Answer Key

1. A
2. A
3. B
4. D
5. B
6. A
7. C
8. A
9. B
10. A
11. D
12. B
13. B
14. B
15. C
16. B
17. D
18. B
19. C
20. D
21. A
22. A
23. C
24. A
25. C
26. B
27. D
28. B
29. A
30. C
31. A
32. B
33. A
34. C
35. B
36. C

Clusters 1-3 Break Down

Apply estimation strategies...

Question 1 ☐

Question 2 ☐

Question 3 ☐

Convert among non-negative fractions...

Question 4 ☐

Question 5 ☐

Question 6 ☐

Compare and order rational numbers

Question 7 ☐

Question 8 ☐

Question 9 ☐

Convert within and between standard...

Question 10 ☐

Question 11 ☐

Question 12 ☐

Solve problems involving ratios...

Question 13 ☐

Question 14 ☐

Question 15 ☐

Question 16 ☐

Solve problems involving proportions

Question 17 ☐

Question 18 ☐

Question 19 ☐

Question 20 ☐

Solve problems involving percentages

Question 21 ☐

Question 22 ☐

Question 23 ☐

Question 24 ☐

Explain the relationship btw var...

Question 25 ☐

Question 26 ☐

Question 27 ☐

Interpret relevant information from....

Question 28 ☐

Question 29 ☐

Question 30 ☐

Evaluate the information in tables...

Question 31 ☐

Question 32 ☐

Question 33 ☐

Calculate geometric quantities

Question 34 ☐

Question 35 ☐

Question 36 ☐

BLANK PAGE

Cluster 4

Perform Arithmetic Operations

with

Rational Numbers

- How to add fractions
- How to add mixed numbers
- How to subtract fractions
- How to subtract mixed numbers
- How to multiply fractions
- How to multiply mixed numbers
- How to divide fractions
- How to divide mixed numbers

Solve Real - World or Multi-Step Problems

with

Rational Numbers

- How to calculate financial adjustments.
- How to calculate net annual salary.
- How to calculate rate of work.
- How to add 3 or more fractions.
- How to perform multiple operations in multi-step word problems.

BLANK PAGE

Perform Arithmetic Operations with Rational Numbers

You will learn:

How to add fractions

How to add mixed numbers

How to subtract fractions

How to subtract mixed numbers

How to multiply fractions

How to multiply mixed numbers

How to divide fractions

How to divide mixed numbers

Study Tips

Read and study EVERY example problem.

Complete EVERY practice problem.

Check to make sure all answers are correct.

Go back to correct the questions you answered incorrectly.

If you don't receive at least an 80% on the review, go back and practice the topic.

Adding & Subtracting Fractions

KISS IT

Step 1: Multiply the numbers at the bottom

Step 2: Cross multiply

Step 3: Add or Subtract

Example 1

$$\frac{1}{4}+\frac{1}{2}$$

Step 1: Multiply the two numbers at the bottom

$$\frac{1}{4} + \frac{1}{2} = \frac{}{\mathbf{8}} + \frac{}{\mathbf{8}}$$

Multiply (4 × 2)

Step 2: Cross Multiply

$$\frac{1}{4} \times \frac{1}{2} = \frac{2}{8} + \frac{4}{8}$$

Multiply (1 × 2)

Multiply (1 × 4)

Step 3: Add

$$\frac{2}{8} + \frac{4}{8} = \frac{6}{8}$$

*Don't forget to reduce your fractions

$$\frac{6 \div 2}{8 \div 2} = \frac{3}{4}$$

Practice 1

1. $\frac{1}{2}+\frac{1}{3}$

2. $\frac{2}{5}+\frac{1}{4}$

3. $\frac{6}{7}+\frac{1}{2}$

4. $\frac{1}{11}+\frac{1}{3}$

5. $\frac{4}{9}+\frac{2}{7}$

6. $\frac{1}{5}+\frac{2}{3}$

7. $\frac{6}{7}+\frac{5}{6}$

8. $\frac{10}{11}+\frac{2}{3}$

9. $\frac{5}{8}+\frac{1}{2}$

10. $\frac{4}{5}+\frac{3}{4}$

11. $\frac{1}{5}+\frac{2}{7}$

12. $\frac{2}{11}+\frac{5}{9}$

13. $\frac{1}{7}+\frac{1}{3}$

14. $\frac{2}{4}+\frac{1}{5}$

15. $\frac{8}{10}+\frac{1}{4}$

16. $\frac{4}{7}+\frac{3}{5}$

17. $\frac{2}{7}+\frac{1}{9}$

18. $\frac{3}{5}+\frac{2}{3}$

19. $\frac{1}{5}+\frac{3}{4}$

20. $\frac{8}{11}+\frac{2}{7}$

Example 2

$$\frac{7}{8}-\frac{2}{3}$$

Step 1: Multiply the two numbers at the bottom

$$\frac{7}{8} \longrightarrow - \frac{2}{3} = \frac{}{24} - \frac{}{24}$$

Multiply (8 × 3)

Step 2: Cross Multiply

$$\frac{7}{8} \times \frac{2}{3} = \frac{21}{24} - \frac{16}{24}$$

Multiply (7 × 3)

Multiply (8 × 2)

Step 3: Subtract

$$\frac{21}{24} - \frac{16}{24} = \frac{5}{24}$$

Practice 2

1. $\frac{1}{2}-\frac{1}{3}$
2. $\frac{2}{5}-\frac{1}{4}$
3. $\frac{6}{7}-\frac{1}{2}$
4. $\frac{1}{3}-\frac{1}{11}$
5. $\frac{4}{9}-\frac{2}{7}$
6. $\frac{2}{3}-\frac{1}{5}$
7. $\frac{6}{7}-\frac{5}{6}$
8. $\frac{10}{11}-\frac{2}{3}$
9. $\frac{5}{8}-\frac{1}{2}$
10. $\frac{4}{5}-\frac{3}{4}$
11. $\frac{9}{10}-\frac{1}{2}$
12. $\frac{2}{5}-\frac{1}{3}$
13. $\frac{2}{3}-\frac{1}{2}$
14. $\frac{4}{5}-\frac{2}{3}$
15. $\frac{8}{9}-\frac{3}{4}$
16. $\frac{10}{12}-\frac{1}{4}$
17. $\frac{2}{4}-\frac{1}{8}$
18. $\frac{2}{3}-\frac{1}{9}$
19. $\frac{3}{5}-\frac{2}{4}$
20. $\frac{6}{7}-\frac{3}{8}$

Adding & Subtracting Mixed Numbers

KISS IT

Step 1: Convert to improper fraction

Step 2: Multiply the numbers at the bottom

Step 3: Cross Multiply

Step 4: Add or Subtract

Example 3

$$2\frac{1}{4}+3\frac{7}{8}$$

Step 1: Convert to improper fraction

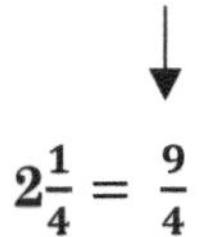

Multiply whole number and bottom number (**2 × 4**)

Add answer from previous step to top number (**8 + 1**)

$2\frac{1}{4} = \frac{9}{4}$

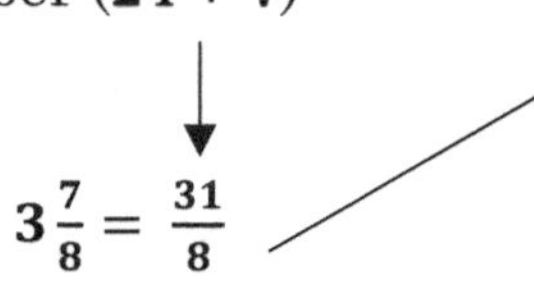

Multiply whole number and bottom number (**3 × 8**)

Add answer from previous step to top number (**24 + 7**)

$3\frac{7}{8} = \frac{31}{8}$

The bottom number will stay the same

Step 2: Multiply the numbers at the bottom

$\frac{9}{4} + \frac{31}{8} = \frac{}{32} + \frac{}{32}$

Multiply (4 × 8)

Step 3: Cross Multiply

$\frac{9}{4} \times \frac{31}{8} = \frac{72}{32} + \frac{124}{32}$

Multiply (9 × 8)

Multiply (31 × 4)

Step 4: Add

$$\frac{72}{32} + \frac{124}{32} = \frac{196}{32}$$

*Don't forget to reduce your fractions

$$\frac{196 \div 4}{32 \div 4} = \frac{49}{8}$$

Improper fractions must be converted to mixed numbers

$$\frac{49}{8} = 6\frac{1}{8}$$

Practice 3

1. $1\frac{1}{2}+2\frac{1}{4}$	6. $3\frac{4}{5}+5\frac{2}{4}$	11. $1\frac{2}{7}+3\frac{2}{5}$	16. $7\frac{2}{3}+1\frac{1}{4}$
2. $5\frac{1}{6}+2\frac{2}{5}$	7. $8\frac{2}{5}+4\frac{3}{4}$	12. $6\frac{2}{5}+5\frac{1}{5}$	17. $11\frac{1}{4}+5\frac{2}{3}$
3. $2\frac{2}{3}+4\frac{1}{8}$	8. $10\frac{1}{2}+4\ \frac{2}{5}$	13. $8\frac{10}{11}+2\frac{1}{2}$	18. $2\frac{1}{8}+3\frac{1}{6}$
4. $5\frac{1}{7}+6\frac{1}{2}$	9. $6\frac{1}{3}+5\frac{3}{5}$	14. $4\frac{1}{6}+3\frac{3}{5}$	19. $4\frac{3}{7}+3\frac{2}{6}$
5. $2\frac{3}{8}+1\frac{1}{6}$	10. $4\frac{1}{7}+5\frac{1}{5}$	15. $1\frac{2}{9}+6\frac{1}{4}$	20. $1\frac{2}{3}+2\frac{1}{4}$

Example 4

$$4\frac{1}{3}-2\frac{3}{4}$$

Step 1: Convert to improper fraction

$4 \rightarrow \frac{1}{3}$

Multiply whole number and bottom number (**4 × 3**)

Add answer from previous step to top number (**12 + 1**)

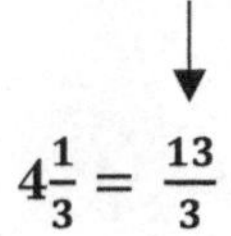

$4\frac{1}{3} = \frac{13}{3}$

$2 \rightarrow \frac{3}{4}$

Multiply whole number and bottom number (**2 × 4**)

Add answer from previous step to top number (**8 + 3**)

$2\frac{3}{4} = \frac{11}{4}$

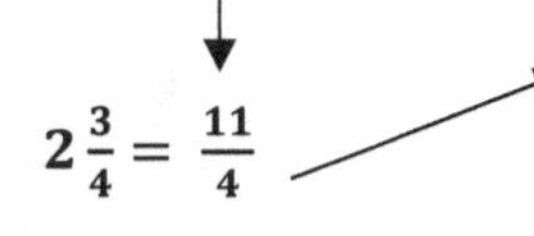

The bottom number will stay the same

Step 2: Multiply the numbers at the bottom

$$\frac{13}{3} - \frac{11}{4} = \frac{\quad}{12} - \frac{\quad}{12}$$

Multiply (3 × 4)

Step 3: Cross Multiply

$$\frac{13}{3} - \frac{11}{4} = \frac{52}{12} - \frac{33}{12}$$

Multiply (13 × 4)

Multiply (11 × 3)

Step 4: Subtract

$$\frac{52}{12} - \frac{33}{12} = \frac{19}{12} \longrightarrow$$

Improper fractions must be converted to mixed numbers

$$\frac{19}{12} = 1\frac{7}{12}$$

Practice 4

1. $2\frac{1}{4} - 1\frac{1}{2}$
2. $5\frac{1}{5} - 2\frac{2}{5}$
3. $4\frac{1}{8} - 1\frac{1}{3}$
4. $6\frac{1}{2} - 4\frac{3}{7}$
5. $2\frac{3}{8} - 1\frac{1}{5}$
6. $5\frac{2}{4} - 2\frac{6}{12}$
7. $8\frac{2}{5} - 4\frac{1}{6}$
8. $10\frac{1}{2} - 2\frac{3}{5}$
9. $6\frac{1}{3} - 5\frac{3}{5}$
10. $5\frac{1}{5} - 4\frac{2}{3}$
11. $3\frac{1}{5} - 2\frac{2}{3}$
12. $1\frac{7}{8} - 1\frac{3}{7}$
13. $7\frac{1}{4} - 4\frac{2}{5}$
14. $9\frac{4}{5} - 3\frac{4}{6}$
15. $12\frac{3}{8} - 9\frac{2}{2}$
16. $4\frac{1}{7} - 2\frac{2}{3}$
17. $5\frac{1}{6} - 4\frac{1}{4}$
18. $8\frac{1}{2} - 4\frac{2}{5}$
19. $7\frac{2}{6} - 1\frac{2}{7}$
20. $9\frac{2}{8} - 4\frac{2}{3}$

Multiplying Fractions

KISS IT

Step 1: Multiply straight across top & bottom

Example 5

$$\frac{1}{4} \times \frac{2}{9}$$

Step 1: Multiply straight across

$$\frac{1}{4} \times \frac{2}{9} = \frac{1 \times 2}{4 \times 9} = \frac{2}{36} = \frac{1}{18}$$

Practice 5

1. $\frac{2}{9} \times \frac{3}{4}$
2. $\frac{2}{5} \times \frac{1}{8}$
3. $\frac{1}{4} \times \frac{7}{8}$
4. $\frac{1}{2} \times \frac{1}{3}$
5. $\frac{2}{3} \times \frac{1}{5}$
6. $\frac{2}{3} \times \frac{7}{9}$
7. $\frac{4}{5} \times \frac{3}{4}$
8. $\frac{1}{4} \times \frac{2}{7}$
9. $\frac{1}{4} \times \frac{3}{8}$
10. $\frac{1}{11} \times \frac{2}{11}$
11. $\frac{6}{8} \times \frac{1}{3}$
12. $\frac{1}{2} \times \frac{3}{4}$
13. $\frac{3}{7} \times \frac{2}{5}$
14. $\frac{3}{5} \times \frac{2}{4}$
15. $\frac{4}{5} \times \frac{2}{7}$
16. $\frac{3}{4} \times \frac{4}{7}$
17. $\frac{2}{3} \times \frac{7}{10}$
18. $\frac{1}{5} \times \frac{1}{3}$
19. $\frac{4}{5} \times \frac{2}{8}$
20. $\frac{2}{3} \times \frac{3}{6}$

Multiplying Mixed Numbers

KISS IT

Step 1: Convert to improper fraction

Step 2: Multiply straight across top & bottom

Example 6

$$1\frac{1}{3} \times 2\frac{2}{4}$$

Step 1: Convert to improper fraction

Multiply whole number and bottom number (**1 × 3**). Add answer from previous step to top number (**3 + 1**)

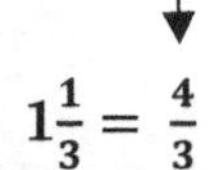

$2 \to \frac{2}{4}$

Multiply whole number and bottom number (**2 × 4**). Add answer from previous step to top number (**8 + 2**)

$$2\frac{2}{4} = \frac{10}{4}$$

The bottom number will stay the same

Step 3: Multiply straight across top & bottom

$$\frac{4}{3} \times \frac{10}{4} = \frac{4 \times 10}{3 \times 4} = \frac{40}{12}$$

*Don't forget to reduce

$$\frac{40 \div 4}{12 \div 4} = \frac{10}{3}$$

Improper fractions must be converted to mixed numbers

$$\frac{10}{3} = 3\frac{1}{3}$$

Practice 6

1. $1\frac{1}{2} \times 2\frac{1}{4}$	**6.** $3\frac{2}{5} \times 1\frac{2}{4}$	**11.** $6\frac{1}{3} \times \frac{2}{3}$	**16.** $\frac{6}{3} \times 1\frac{3}{4}$
2. $3\frac{1}{5} \times 1\frac{2}{3}$	**7.** $8\frac{1}{4} \times \frac{1}{5}$	**12.** $2\frac{3}{5} \times \frac{1}{3}$	**17.** $3\frac{4}{5} \times \frac{7}{4}$
3. $1\frac{2}{5} \times 4\frac{1}{3}$	**8.** $2\frac{3}{5} \times 1\frac{5}{6}$	**13.** $5\frac{1}{3} \times 5\frac{2}{4}$	**18.** $7\frac{1}{2} \times \frac{3}{2}$
4. $4\frac{1}{2} \times \frac{2}{3}$	**9.** $8\frac{2}{7} \times 1\frac{1}{7}$	**14.** $1\frac{3}{4} \times 1\frac{2}{4}$	**19.** $\frac{7}{6} \times \frac{6}{7}$
5. $\frac{5}{6} \times 3\frac{1}{3}$	**10.** $1\frac{2}{7} \times 1\frac{2}{8}$	**15.** $\frac{6}{5} \times \frac{9}{7}$	**20.** $2\frac{3}{7} \times 2\frac{1}{5}$

Dividing Fractions

KISS IT

Step 1: Change multiplication to division

Step 2: Flip second fraction

Step 3: Multiply straight across

Example 7

$$\frac{2}{3} \div \frac{3}{4}$$

Step 1: Change multiplication to division

$$\frac{2}{3} \times \frac{3}{4}$$

Step 2: Flip second fraction

$$\frac{2}{3} \times \frac{4}{3}$$

Step 3: Multiply straight across

$$\frac{2}{3} \times \frac{4}{3} = \frac{2 \times 4}{3 \times 3} = \frac{8}{9}$$

Practice 7

1. $\frac{2}{9} \div \frac{3}{4}$
2. $\frac{2}{5} \div \frac{1}{8}$
3. $\frac{1}{4} \div \frac{7}{8}$
4. $\frac{1}{2} \div \frac{1}{3}$
5. $\frac{2}{3} \div \frac{1}{5}$
6. $\frac{2}{3} \div \frac{7}{9}$
7. $\frac{4}{5} \div \frac{3}{4}$
8. $\frac{1}{4} \div \frac{2}{7}$
9. $\frac{1}{4} \div \frac{3}{8}$
10. $\frac{1}{11} \div \frac{2}{11}$
11. $\frac{6}{8} \div \frac{1}{3}$
12. $\frac{1}{2} \div \frac{3}{4}$
13. $\frac{3}{7} \div \frac{2}{5}$
14. $\frac{3}{5} \div \frac{2}{4}$
15. $\frac{4}{5} \div \frac{2}{7}$
16. $\frac{3}{4} \div \frac{4}{7}$
17. $\frac{2}{3} \div \frac{7}{10}$
18. $\frac{1}{5} \div \frac{1}{3}$
19. $\frac{4}{5} \div \frac{2}{8}$
20. $\frac{2}{3} \div \frac{3}{6}$

Dividing Mixed Numbers

KISS IT

Step 1: Convert to improper fraction

Step 2: Change division to multiplication

Step 3: Flip second fraction

Step 4: Multiply straight across

Example 8

$$1\frac{1}{3} \div 2\frac{2}{4}$$

Step 1: Convert to improper fraction

Multiply whole number and bottom number (**1 × 3**). Add answer from previous step to top number (**3 + 1**)

$$1\frac{1}{3} = \frac{4}{3}$$

$2 \longrightarrow \frac{2}{4}$

Multiply whole number and bottom number (**2 × 4**). Add answer from previous step to top number (**8 + 2**)

$$2\frac{2}{4} = \frac{10}{4}$$

The bottom number will stay the same

Step 2: Change division to multiplication $\frac{4}{3} \times \frac{10}{4}$

Step 3: Flip second fraction $\frac{4}{3} \times \frac{4}{10}$

Step 4: Multiply straight across $\frac{4}{3} \times \frac{4}{10} = \frac{4 \times 4}{3 \times 10} = \frac{16}{30}$

*Don't forget to reduce your fraction

$$\frac{16 \div 2}{30 \div 2} = \frac{8}{15}$$

Practice 8

1. $1\frac{1}{2} \div 2\frac{1}{4}$

2. $3\frac{1}{5} \div 1\frac{2}{3}$

3. $1\frac{2}{5} \div 4\frac{1}{3}$

4. $4\frac{1}{2} \div \frac{2}{3}$

5. $\frac{5}{6} \div 3\frac{1}{3}$

6. $3\frac{2}{5} \div 1\frac{2}{4}$

7. $8\frac{1}{4} \div \frac{1}{5}$

8. $2\frac{3}{5} \div 1\frac{5}{6}$

9. $8\frac{2}{7} \div 1\frac{1}{7}$

10. $1\frac{2}{7} \div 1\frac{2}{8}$

11. $6\frac{1}{3} \div \frac{2}{3}$

12. $2\frac{3}{5} \div \frac{1}{3}$

13. $5\frac{1}{3} \div 5\frac{2}{4}$

14. $1\frac{3}{4} \div 1\frac{2}{4}$

15. $\frac{6}{5} \div \frac{9}{7}$

16. $\frac{6}{3} \div 1\frac{3}{4}$

17. $3\frac{4}{5} \div \frac{7}{4}$

18. $7\frac{1}{2} \div \frac{3}{2}$

19. $\frac{7}{6} \div \frac{6}{7}$

20. $2\frac{3}{7} \div 2\frac{1}{5}$

Answer Key

Practice 1

1. $\frac{5}{6}$	11. $\frac{17}{35}$
2. $\frac{13}{20}$	12. $\frac{73}{99}$
3. $1\frac{5}{14}$	13. $\frac{10}{21}$
4. $\frac{14}{33}$	14. $\frac{7}{10}$
5. $\frac{46}{63}$	15. $1\frac{1}{20}$
6. $\frac{13}{15}$	16. $1\frac{6}{35}$
7. $1\frac{29}{42}$	17. $\frac{25}{63}$
8. $1\frac{19}{33}$	18. $1\frac{4}{15}$
9. $1\frac{1}{8}$	19. $\frac{19}{20}$
10. $1\frac{11}{20}$	20. $1\frac{1}{77}$

Practice 2

1. $\frac{1}{6}$	11. $\frac{2}{5}$
2. $\frac{3}{20}$	12. $\frac{1}{15}$
3. $\frac{5}{14}$	13. $\frac{1}{6}$
4. $\frac{8}{33}$	14. $\frac{2}{15}$
5. $\frac{10}{63}$	15. $\frac{5}{36}$
6. $\frac{7}{15}$	16. $\frac{7}{12}$
7. $\frac{1}{42}$	17. $\frac{3}{8}$
8. $\frac{8}{33}$	18. $\frac{5}{9}$
9. $\frac{1}{8}$	19. $\frac{1}{10}$
10. $\frac{1}{20}$	20. $\frac{27}{56}$

Practice 3

1. $3\frac{3}{4}$	11. $4\frac{24}{35}$
2. $7\frac{17}{30}$	12. $11\frac{3}{5}$
3. $6\frac{19}{24}$	13. $11\frac{9}{22}$
4. $11\frac{9}{14}$	14. $7\frac{23}{30}$
5. $3\frac{13}{24}$	15. $7\frac{17}{36}$
6. $9\frac{3}{10}$	16. $8\frac{11}{12}$
7. $13\frac{3}{20}$	17. $16\frac{11}{12}$
8. $14\frac{9}{10}$	18. $5\frac{7}{24}$
9. $11\frac{14}{15}$	19. $7\frac{16}{21}$
10. $9\frac{12}{35}$	20. $3\frac{11}{12}$

Practice 4

1. $\frac{3}{4}$	11. $\frac{8}{15}$
2. $2\frac{4}{5}$	12. $\frac{25}{56}$
3. $2\frac{19}{24}$	13. $2\frac{17}{20}$
4. $2\frac{1}{14}$	14. $6\frac{2}{15}$
5. $1\frac{7}{40}$	15. $2\frac{3}{8}$
6. 3	16. $1\frac{10}{21}$
7. $4\frac{7}{30}$	17. $\frac{11}{12}$
8. $7\frac{9}{10}$	18. $4\frac{1}{10}$
9. $\frac{11}{15}$	19. $6\frac{1}{21}$
10. $\frac{8}{15}$	20. $4\frac{7}{12}$

Practice 5

1. $\frac{1}{6}$	11. $\frac{1}{4}$
2. $\frac{1}{20}$	12. $\frac{3}{8}$
3. $\frac{7}{32}$	13. $\frac{6}{35}$
4. $\frac{1}{6}$	14. $\frac{3}{10}$
5. $\frac{2}{15}$	15. $\frac{8}{35}$
6. $\frac{14}{27}$	16. $\frac{3}{7}$
7. $\frac{3}{5}$	17. $\frac{7}{15}$
8. $\frac{1}{14}$	18. $\frac{1}{15}$
9. $\frac{3}{32}$	19. $\frac{1}{5}$
10. $\frac{2}{121}$	20. $\frac{1}{3}$

Practice 6

1. $3\frac{3}{8}$	11. $4\frac{2}{9}$
2. $5\frac{1}{3}$	12. $\frac{13}{15}$
3. $6\frac{1}{15}$	13. $29\frac{1}{3}$
4. 3	14. $2\frac{5}{8}$
5. $2\frac{7}{9}$	15. $1\frac{19}{35}$
6. $5\frac{1}{10}$	16. $3\frac{1}{2}$
7. $1\frac{13}{20}$	17. $6\frac{13}{20}$
8. $4\frac{23}{30}$	18. $11\frac{1}{4}$
9. $9\frac{23}{49}$	19. 1
10. $1\frac{17}{28}$	20. $5\frac{12}{35}$

Practice 7

1. $\frac{8}{27}$	11. $2\frac{1}{4}$
2. $3\frac{1}{5}$	12. $\frac{2}{3}$
3. $\frac{2}{7}$	13. $1\frac{1}{14}$
4. $1\frac{1}{2}$	14. $1\frac{1}{5}$
5. $3\frac{1}{3}$	15. $2\frac{4}{5}$
6. $\frac{6}{7}$	16. $1\frac{5}{16}$
7. $1\frac{1}{15}$	17. $\frac{20}{21}$
8. $\frac{7}{8}$	18. $\frac{3}{5}$
9. $\frac{2}{3}$	19. $3\frac{1}{5}$
10. $\frac{1}{2}$	20. $1\frac{1}{3}$

Practice 8

1. $\frac{2}{3}$	11. $9\frac{1}{2}$
2. $1\frac{23}{25}$	12. $7\frac{4}{5}$
3. $\frac{21}{65}$	13. $\frac{32}{33}$
4. $6\frac{3}{4}$	14. $1\frac{1}{6}$
5. $\frac{1}{4}$	15. $\frac{14}{15}$
6. $2\frac{4}{15}$	16. $1\frac{1}{7}$
7. $41\frac{1}{4}$	17. $2\frac{6}{35}$
8. $1\frac{23}{55}$	18. 5
9. $7\frac{1}{4}$	19. $1\frac{13}{36}$
10. $1\frac{1}{35}$	20. $1\frac{8}{77}$

Solve Real – World One Or Multi-Step Problems With Rational Numbers

You will learn:

How to calculate financial adjustments

How to calculate net annual salary

How to calculate rate of work

How to add 3 or more fractions

How to perform multiple operations in multi-step word problems

Study Tips

Read and study EVERY example problem.

Complete EVERY practice problem.

Check to make sure all answers are correct.

Go back to correct the questions you answered incorrectly.

If you don't receive at least an 80% on the review, go back and practice the topic.

Financial Adjustments

KISS IT

Step 1: Identify key words

Step 2: Perform the given operation(s)

Example 1

Last Friday, Malik's bank account had a balance of $2,067.59. Here's a list of his account activity since then. What is his new balance after his last deposit on Thursday?

Saturday	Deposited a check for $506.92
Monday	wrote a check for $93.25
Tuesday	Charged a $35 bank fee
Thursday	Deposited $150

Step 1: Identify key words

Saturday	**Deposited** a check for $506.92
Monday	**wrote a check** for $93.25
Tuesday	**Charged** a $35 bank **fee**
Thursday	**Deposited** $150

***Transactions related to bank accounts primarily deal with addition or subtraction**

- Deposit: addition
- Write a check: subtraction
- Fee: subtraction
- Deposit: addition

Step 2: Perform the given operations

(Starting balance) + (Deposit) – (check) – (bank fee) + (deposit)

$2,067.59 + $506.92 – $93.25 – $35 + $150 = **$2,596.26**

Answer: $2,596.26

Practice 1

1. On Monday Kevon's available balance was $3,489.92. Tuesday, he made a purchase at the local supermarket for $72.34. Wednesday, he wrote a check for his personal trainer totaling $300. Thursday, his paycheck ($2,389.76) was deposited. What was his available balance on Friday?

2. On Saturday Aaliyah's available balance was $345.22. Sunday, she made a deposit of $45. Monday, she wrote a check for her brother totaling $92. What was her available balance on Tuesday?

3. Sandra spent $235.67 on beauty supplies, wrote a check to the electric company for $89.62, and deposited a check for $500. If her starting balance was $300, and her bank charges an overdraft fee of $50, what is her current balance?

4. Brian's checking account had an available balance of $10,235.54 on Wednesday. That same day, he made a purchase of $898.21 at a local boutique. Thursday, his bank charged him a monthly fee of $32. He also wrote a check for his daughter that same day for $500. What was his available balance Friday?

5. Gloria's checking account balance was $3,500 on Monday, and by Friday it was $5000. If she wrote a check for $30, swiped her card for $800, and made an atm withdrawal of $700, how much money did she deposit?

Example 2

Janet receives a paycheck bi-monthly for $3,625.67. She pays $325.67 in taxes, $150 for health insurance, $50 for dental, and $300 to her savings each pay period. What is her net pay each month?

Step 1: Identify key words

Janet receives a pay check **bi-monthly** for $3,625.67. She **pays** $325.67 in taxes, $150 for health insurance, $50 for dental, and $300 **to her savings** each pay period. What is her **net pay** each month?

- Bi-monthly: twice per month/every two weeks
- Net pay: Paycheck – deductions

Step 2: Perform the given operations

$3,625.67 – $325.67 – $150 – $50 – $300 = $2,800 **bi-weekly**

$2800 × 2 = $5,600 **monthly**

Answer: $5,600

Practice 2

1. Teshia receives a paycheck bi-monthly for $6,758.92. She pays $2,456.67 in taxes, $500 for health insurance, $135 for dental, and $1500 to her savings each pay period. What is her net pay each month?

2. Tommy receives a monthly paycheck for $3,879.32. He pays $398.21 in taxes each pay period. What is his net pay each month?

3. Workers at the local plant are paid $1000 each week. They pay $200 in taxes, $25 for health insurance, and $15 for dental every two weeks. What is their net pay each month?

4. Whitney is paid $2500 every two weeks at her first job. Her second job pays her $2000 monthly. If she pays a total of $1500 in taxes for both jobs each month, what is her net pay each month?

5. Lance is paid $300 for each event he shoots. For the month of June he is booked for 27 events. If he pays $900 in taxes, has $500 sent to his savings account, what is his net pay or June?

Example 3

Kevin is paid $4,325 bi-monthly, and $346 is withheld from each paycheck. How much is his net annual salary?

Step 1: Identify key words

Kevin is paid $4,325 **bi-monthly**, and $346 is **withheld from each paycheck.** How much is his **net annual salary**?

- Withheld from each paycheck: Deduction
- Bi-monthly: twice per month/ every two weeks
- Net Annual Salary: net pay (paycheck – deductions) × 12

Step 2: Perform the given operations

Net Pay: ($4,325 × 2) – ($346 × 2) = $7,958

Net Annual Salary: $7,958 × 12 = **$95,496**

Answer: $95,496

Practice 3

1. Alexis is paid $1,245 every two weeks. If $137 is withheld from each paycheck, what is her net annual salary?

2. Kayla is paid $8,921 each month. If $536 is withheld from each paycheck (bi-monthly), what is her net annual salary?

3. Robert is paid $889.23 bi-monthly. If $236.78 is withheld from his paychecks each month, what is his net annual salary?

4. Destiny receives a paycheck each month totaling $3,765.90. If $543.98 is withheld from her paycheck each month, what is her net annual salary?

5. Anthony receives a paycheck each month for $2000. If $234.56 is withheld from each paycheck, what is his net annual salary?

6. David's job issues a paycheck once each month for $1500. If $50 is deducted every two weeks for dental insurance, what is his net annual salary?

7. Joanne receives a quarterly check from Amazon for her famous recipe book for $3,500. If $500 is taken from each check for taxes, what is her net annual salary?

8. If Amanda works 40 hours, her paycheck is $2,500 each week. If $250 in deductions is taken from each check, what is her net annual salary?

9. Jacob takes home $5,798 each month after deductions. What is his net annual salary?

Rate of work

KISS IT

Step 1: Multiply the two values

Step 2: Add the two values

Step 3: Divide (step one) by (step two)

Example 4

Alexis needed 4 hours to complete her exam, and Shelly needed 6 hours to complete the same exam. If they were given the opportunity to work together, how long would it take them to complete the exam?

Step 1: Multiply the two values

$4 \times 6 = 24$

Step 2: Add the two values

$4 + 6 = 10$

Step 3: Divide the values

$24 \div 10 = \mathbf{2.4}$

Answer: 2.4 hours

Example 5

Sabrina can finish a painting in $3\frac{1}{2}$ hours, and Kevin can complete a painting in $1\frac{1}{2}$ hours. If they work together, how long would it take them to complete a painting?

Step 1: Multiply the two values

$3\frac{1}{2} \times 1\frac{1}{2} = 5\frac{1}{4}$

Step 2: Add the two values

$3\frac{1}{2} + 1\frac{1}{2} = 5$

Step 3: Divide the values

$5\frac{1}{4} \div 5 = 1\frac{1}{20}$

Answer: $1\frac{1}{20}$ hours

Practice 4

1. It took Ms. Senat 3 hours to finalize the final exam grades for her students. Ms. Perriman completed her grades in 4 hours. If they worked together, how long would it take them to complete the grades?

2. Sarah finished her senior paper in 3 months, and David finished his in 5 months. If they worked together, how long would it take for them to complete the paper?

3. Mark can mow a lawn in 2 hours, and Mathew can mow a lawn in 3 hours. If they work together how long will it take them to mow a lawn?

4. Steve can wash dishes in 30 minutes, and Mary can wash dishes in 15 minutes. If they work together, how long will it take them to wash dishes?

5. Kevon painted his room in 4 hours, and Kevin painted his room in 2 hours. If they work together how long will it take them to paint one room?

6. Gloria can bake 10 cakes in 2 hours, and Grace can back 10 cakes in 2.5 hours. If they work together, how long will it take them to bake 10 cakes together?

7. Emmanuel can write 4 poems in 2 hours, and Jakara can write the same number of poems in 5 hours. If they work together how long will it take them to complete 4 poems?

8. It takes Allen 45 minutes to wash dishes, it takes Rose 30 minutes to complete the same task. If they work together how long will it take them to complete the dishes?

9. David cleans the kitchen in 30 minutes flat. He hired a new housekeeper who can complete the job in 20 minutes. If they work together, how long will it take them to clean the kitchen?

10. George can paint one wall in 13 minutes. Gina can paint the same size wall in 27 minutes. If they work together, how long will it take them to paint?

11. Stephanie can complete her weekly lesson planning in two hours. Brittney can complete her lesson planning in 1 hour. If they work together, how long will it take them to complete the task?

12. Adrianne can complete a full head of rope twists in 6 hours. Michelle can complete the same task in 4 hours. If they work together how long will it take them to finish one customer?

13. Kenny can build 4 partitions in 8 hours. Michael can build the same number of partitions in 10 hours. If they work together, how long will it take them to complete the task?

14. Anthony can wash 10 cars in 3 hours. Lee can wash 10 cars in $2\frac{1}{2}$ hours. If they work together, how long will it take them to complete the task?

15. Ashton can register 15 patients in 4.5 hours. Layla can register the same number of patients in 3 hours. If they work together, how long will it take them to complete the task?

Review

1. Alexis has a jump rope that is 12 meters long. If her brother cuts $\frac{3}{2}$ meters and her sister cuts $\frac{4}{5}$ meters from the rope, what is the length of the remaining rope?

a. 13 b. $14\frac{3}{10}$ c. $9\frac{7}{10}$ d. 11

2. Robert purchased a container of assorted fruit including $\frac{1}{2}$ lb. of strawberries, $\frac{2}{3}$ lb. of grapes, $\frac{3}{5}$ lb. of apples, and $\frac{3}{8}$ lb. of pineapples. What is the total weight of the fruit in the container?

a. $\frac{1}{2}$ b. $2\frac{17}{120}$ c. $\frac{9}{18}$ d. $2\frac{10}{120}$

3. One bag of potatoes costs $3.69. If Courtney needs 8 bags for the restaurant, how much will she spend?

a. $0.46 b. $29.52 c. $2.17 d. $29.25

4. If Kasey eats $\frac{1}{5}$ of a pizza, Corey eats $\frac{2}{9}$ of a pizza and James eats $\frac{1}{3}$ of a pizza, how much pizza did they eat altogether?

a. $\frac{34}{45}$ b. $\frac{45}{34}$ c. $\frac{4}{17}$ d. . $\frac{17}{4}$

5. Allen purchased three bags of fresh seafood for his annual family cookout. The first bag was filled with shrimp weighing $17\frac{1}{2}$ lbs., the second bag was filled with oysters weighing $12\frac{3}{4}$ lbs., and the third bag was filled with lobster tails weighing $7\frac{1}{4}$ lbs. What is the total weight of the three bags of food?

a. $36\frac{11}{20}$ lbs. b. $37\frac{7}{20}$ lbs. c. $37\frac{1}{2}$ lbs. d. $35\frac{1}{20}$ lbs.

6. A lecture hall can hold 225 students. If each row holds 15 students, how many rows are there?

a. 3,375 b. 10 c. 3,300 d. 15

7. Alyssa and her sister baked 100 cookies for the school's annual bake sale. If each cookie was sold for 75 cents, how much money would they make if they sold 20% of the cookies they baked?

a. $75 b. $1500 c. $60 d. $15

8. Daisy receives a monthly check for $3500, and $375 is withheld from each paycheck. What is her net annual salary?

a. $37,500 b. $42,000 c. $4,500 d. $46,500

9. Sunny can paint a small room in 1.5 hours, and Billy can paint the same size room in 2 hours. If they worked together, how long will it take for them to paint a room?

a. 3 hours b. 3.5 hours c. .86 hours d. 1.2 hours

10. The local car lot has 74 rows of cars, if there are 3,330 total cars how many cars are in each row?

a. 45 b. 3,404 c. 246,420 d. 3,256

11. Lindsey's favorite roll of wrapping paper is 10 feet long. If she uses 3.25 feet for her mom's gift and $2\frac{1}{8}$ feet for her daughter's gift, how much wrapping paper will she have left?

a. $5\frac{3}{8}$ b. $6\frac{3}{4}$ c. $4\frac{5}{8}$ d. $15\frac{3}{8}$

12. Ashley is paid $3,452 each month. If $236 is withheld every two weeks, what is her net annual salary?

a. $41,424 b. $2,832 c. $35,760 d. $44,256

13. On Monday Kevon's available balance was $489.92. Tuesday, he made a purchase at the supermarket for $52.84. Wednesday, he wrote a check for his personal trainer totaling $150. Thursday, his paycheck ($689.76) was deposited. What was his available balance on Friday?

a. $1,382.52 b. $976.84 c. $689.76 d. $1,276.84

14. It takes student A 29 minutes to complete the math worksheet, and student B 45 minutes to complete the same task. If they work together, how long will it take them to complete the worksheet?

a. 37 mins b. .06 mins c. 8 mins d. 17.6 mins

15. Allen is 5 feet 9 inches tall, Terrel is 6 feet 2 inches tall, and Stump is 5 feet 5 inches tall. What is the average height of this group?

a. $5\frac{7}{9}$ feet b. $17\frac{6}{10}$ feet c. $17\frac{1}{3}$ feet d. $5\frac{1}{3}$ feet

16. Alex cut a piece of tape into 5 random lengths: $1\frac{3}{4}$ in, 2.3 in, $1\frac{2}{3}$ in, $\frac{7}{8}$ in, and $\frac{1}{2}$ in. What is the total length of the original piece of tape she cut?

a. $7\frac{11}{120}$ b. $\frac{11}{120}$ c. $5\frac{11}{170}$ d. $\frac{11}{170}$

17. Dexter receives a paycheck each month totaling $5,985.90. If $904.87 is withheld from his paycheck each month, what is his net annual salary?

a. $60,972.36 b. $71,830.80 c. $82,689.24 d. $60,279.36

18. Monday Siri ate $1\frac{1}{4}$ lbs. of food, Tuesday she ate $1\frac{2}{3}$ lbs. of food, Wednesday she ate $\frac{7}{8}$ lbs. of food, and Thursday she ate $\frac{1}{2}$ lbs. If she ate a total of 5 lbs. of food that week, how much did she eat on Friday?

a. $4\frac{7}{24}$ b. $\frac{17}{24}$ c. $9\frac{7}{24}$ d. $\frac{23}{24}$

19. There are 500 students at the local middle school. If each classroom holds 25 students how many classrooms are there?

a. 12,500 b. 20 c. 475 d. 30

20. Adrianne receives a paycheck for $5,982.87 every two weeks. Each pay period $1,489.76 is withheld from her check. What is her net annual salary?

a.$35,794.44 b. $17,877.12 c. $107,834.64 d. $89,671.56

Answer Key

Practice 1

1. $5,507.34

2. $298.22

3. $424.71

4. $8,805.33

5. $3,030

Practice 2

1. 4,334.50

2. $3,481.11

3. $3,520

4. $5,500

5. $6,700

Practice 3

1. $26,592

2. $94,188

3. $18,500.16

4. $38,663.04

5. $21,185.28

6. $16,800

7. $12,000

8. $108,000

9. $69,576

Practice 4

1. 1.7 hours

2. 1.9 months

3. 1.2 hours

4. 10 minutes

5. 1.3 hours

6. 1.1 hours

7. 1.4 hours

8. 18 mins

9. 12 mins

10. 8.8 mins

11. 0.67 hours

12. 2.4 hours

13. 4.4 hours

14. 1.4 hours

15. 1.8 hours

Review

1. C

2. B

3. B

4. A

5. C

6. D

7. D

8. A

9. C

10. A

11. C

12. C

13. B

14. D

15. A

16. A

17. A

18. B

19. B

20. C

Cluster 4 Review

You should know:

☐ How to add fractions

☐ How to add mixed numbers

☐ How to subtract fractions

☐ How to subtract mixed numbers

☐ How to multiply fractions

☐ How to multiply mixed numbers

☐ How to divide fractions

☐ How to divide mixed numbers

☐ How to calculate financial adjustments

☐ How to calculate net annual salary

☐ How to calculate rate of work

☐ How to add 3 or more fractions

☐ Perform multiple operations in multi-step word problems

Cluster 4 Review

1. $2\frac{1}{8} \times 4\frac{2}{3}$

a. $8\frac{1}{12}$　b. $9\frac{11}{12}$　c. $\frac{12}{119}$　d. $8\frac{3}{16}$

2. $4\frac{1}{9} \div 1\frac{2}{7}$

a. $4\frac{7}{18}$　b. $3\frac{16}{81}$　c. $5\frac{2}{7}$　d. $3\frac{2}{7}$

3. Luke can type a 10-page paper in 12 hours, Cindy can type the same number of pages in 8 hours. If they work together, how long will it take them to complete the paper?

a. 9 hours　b. .21 hours　c. 5.5 hours　d. 4.8 hours

4. Ashley is paid $2,754.89 bi-monthly. If $432.58 in taxes is taken with each paycheck, what is her net annual salary?

a. $55,735.44　b. $33,058.68　c. $27,867.72　d. $76,499.28

5. $6\frac{1}{4} + 1\frac{1}{2}$

a. $7\frac{1}{6}$　b. $7\frac{3}{4}$　c. $7\frac{1}{8}$　d. $5\frac{1}{2}$

6. Terrell caught 5 fishes of various weights: $4\frac{1}{2}$ lbs., 2.3 lbs., $1\frac{7}{8}$ lbs., $6\frac{3}{4}$ lbs., 1.25 lbs. What is the total combined weight of the fish?

a. 14 lbs.　b. $13\frac{10}{14}$ lbs.　c. $16\frac{27}{40}$ lbs.　d. 16 lbs.

7. $\frac{9}{8} - \frac{7}{11}$

a. $\frac{2}{3}$　b. $\frac{16}{19}$　c. $1\frac{67}{88}$　d. $\frac{43}{88}$

8. Maxine weighs 175 lbs., Alexis weighs 145 lbs., Trent weighs 224 lbs., and Larry weighs 295 lbs. If their weight accounts for one-half of the elevator's maximum weight allotted, what is the maximum weight the elevator can carry?

a. 1678 lbs.　b. 839 lbs.　c. 419.50 lbs.　d. 1680 lbs.

9. $\frac{3}{7} + \frac{2}{3}$

a. $1\frac{2}{21}$　b. $\frac{5}{21}$　c. $\frac{21}{23}$　d. $1\frac{6}{21}$

10. An auditorium has enough seats to hold 1,200 students. If there are 20 seats in each row of the auditorium, how many rows are there?

a. 2,400　b. 60　c. 600　d. 24,000

11. $\frac{1}{2} \times \frac{2}{3} \div \frac{2}{9}$

a. $1\frac{1}{2}$ b. $\frac{4}{54}$ c. $\frac{2}{3}$ d. $1\frac{5}{14}$

12. $4\frac{2}{5} - 1\frac{3}{4}$

a. $3\frac{1}{4}$ b. $5\frac{7}{20}$ c. $3\frac{7}{20}$ d. $2\frac{13}{20}$

13. Kendrick's checking account had an available balance of $20,235.54 on Wednesday. That same day, he made a purchase for $8,298.21. Thursday, his bank charged him a monthly fee of $25. He wrote a check the next day for $1000. What was his available balance Friday?

a. $27,508.75 b. $12,962.33 c. $10,912.33 d. $11,912.33

14. $\frac{3}{4} \times \frac{7}{11}$

a. $\frac{2}{3}$ b. $\frac{28}{33}$ c. $\frac{21}{44}$ d. $\frac{3}{4}$

15. Alisha is paid $800 weekly. If $72 is withheld from each paycheck, what is her net annual salary?

a. $9,600 b. $10,464 c. $34,944 d. $41,856

16. Janean can bake and decorate a cake in $2\frac{1}{4}$ hours, Lindsey can bake and decorate a cake in $2\frac{3}{5}$ hours. If they work together, how long will it take them to complete this task?

a. $2\frac{17}{40}$ hours b. $\frac{97}{117}$ hours c. $\frac{20}{97}$ hours d. $1\frac{20}{97}$ hours

17. $\frac{4}{5} \div \frac{1}{2}$

a. $\frac{5}{7}$ b. $1\frac{3}{5}$ c. $\frac{2}{5}$ d. $1\frac{2}{5}$

18. $\frac{3}{5} + \frac{1}{3} + \frac{2}{4}$

a. $1\frac{13}{30}$ b. $\frac{1}{2}$ c. $1\frac{1}{10}$ d. $\frac{13}{30}$

19. Karoline, Edward, and Kimberly each ate $\frac{1}{7}$, $\frac{2}{4}$, and $\frac{1}{3}$ of the sweet potato pie respectively. How much pie is left over?

a. $\frac{1}{42}$ b. $\frac{2}{7}$ c. $\frac{41}{42}$ d. $\frac{5}{7}$

20. At the local supermarket, oranges are $1.29/lb., apples are 75 cents/lb., and bananas are 30 cents/lb. If Ana purchases 3 lbs. of oranges, 2 lbs. of apples, and 5 lbs. of bananas, what is her total?

a. $303.87 b. $ $2.34 c.$10 d. $6.87

Answer Key

1. B

2. B

3. D

4. A

5. B

6. C

7. D

8. A

9. A

10. B

11. A

12. D

13. C

14. C

15. C

16. D

17. B

18. A

19. A

20. D

Cluster 4 Review Break Down

Perform Arithmetic Operations with Rational Numbers

Question 1 ☐

Question 2 ☐

Question 5 ☐

Question 7 ☐

Question 9 ☐

Question 11 ☐

Question 12 ☐

Question 14 ☐

Question 17 ☐

Question 18 ☐

Solve Real - World or Multi-Step Problems with Rational Numbers

Question 3 ☐

Question 4 ☐

Question 6 ☐

Question 8 ☐

Question 10 ☐

Question 13 ☐

Question 15 ☐

Question 16 ☐

Question 19 ☐

Question 20 ☐

BLANK PAGE

Cluster 5

TRANSLATE PHRASES AND SENTENCES INTO EXPRESSIONS EQUATIONS AND INEQUALITIES

- How to translate sentences into expressions
- How to translate sentences into equations
- How to translate sentences into inequalities
- How to create equations from word problems

SOLVE EQUATIONS IN ONE VARIABLE

- How solve equations with variables on both sides
- How to solve equations using the distributive property
- How to solve inequalities
- How to solve absolute value equations
- How to solve absolute value inequalities

BLANK PAGE

Translate Phrases and Sentences Into Expressions Equations and Inequalities

You will learn:

How to translate sentences into expressions

How to translate sentences into equations

How to translate sentences into inequalities

How to create equations from word problems

Study Tips

Read and study EVERY example problem.

Complete EVERY practice problem.

Check to make sure all answers are correct.

Go back to correct the questions you answered incorrectly.

If you don't receive at least an 80% on the review, go back and practice the topic.

Translate Phrases and Sentences into Expressions

KISS IT

Step 1: Identify key words

Step 2: Plug in values

Addition	Subtraction
Altogether In all Total Sum Both Combined How much More than Gain	Difference Change in Fewer than How many more How much more Minus Reduce Less than Loss
Multiplication	**Division**
Of Product Times By Twice Double Triple Each Percent of Fraction of	Each group has Half Separated Equal Sharing Cut up As much Per Divided by

Example 1

Six more than 3 times x

Step 1: Identify key words

Six **more than** 3 **times** a

↓ addition ↓ multiplication

Step 2: Plug in values

$6 + 3 \times a$

$\mathbf{6 + 3a}$

Example 2

Five less than one-third of the sum of x and y

Step 1: Identify key words

Five **less than** one-third **of** the **sum** of x and y

↓ subtraction ↓ multiplication ↘ addition

Note: Subtraction should always be placed at the end of the expression

Step 2: Plug in values $\frac{1}{3}(x + y) - 5$

Practice 1

1. Ten less than 3 divided by x

2. Four more than 5 times c

3. One-half of the sum of 5 and b

4. One-fourth of x plus one-eighth of y

5. Two more than the difference between b and c

6. Seven less than a times b

7. Five more than 2 times z

8. Six less than 4 times t

9. Three times x minus twelve

10. Nine less than half of x

11. Twice z

12. One fourth of the sum of a and 8

13. Four times d plus 19

14. Seven more than 5 times g

15. 10 fewer than y

16. 6 times x plus 90

17. Thirty-six less than 6 times d

18. 5 more than double c

19. Triple b

20. Five less than 3 times x

Translate Phrases and Sentences into Equations

KISS IT

Step 1: Identify key words

Step 2: Plug in values

Equal Sign
is equals equal to is equal to the same as yields

Example 3

Six less than four times b is twenty-five

Step 1: Identify key words

Six **less than** four **times** b **is** twenty-five

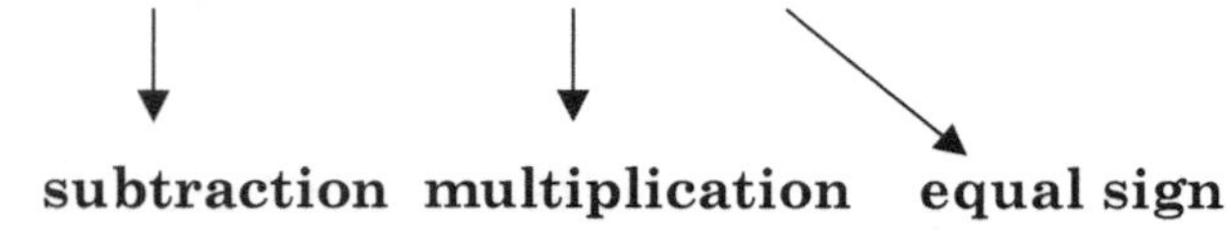

Step 2: Plug in values

Six **less than** four **times** b **is** twenty-five

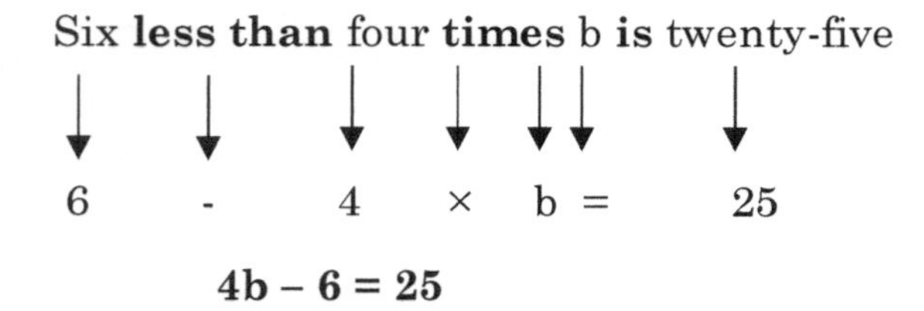

Note: Subtraction should always be placed at the end of the expression

$\mathbf{4b - 6 = 25}$

Example 4

The sum of three times a and four times b is the same as y

Step 1: Identify the key words

The **sum** of three **times** a and four **times** b **is the same as** y

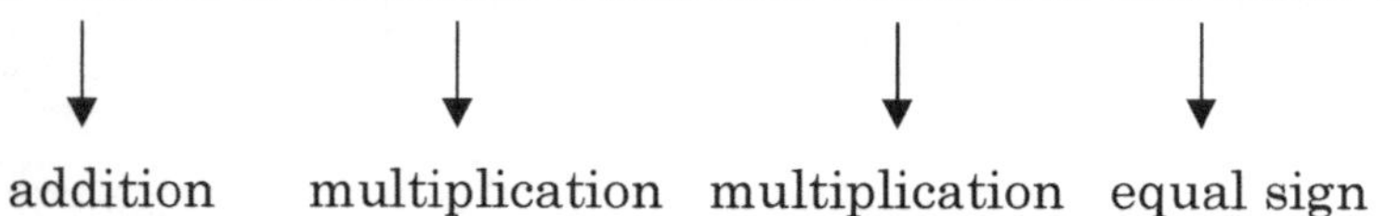

Step 2: Plug in values

$3 \times a + 4 \times b = y$

$3a + 4b = y$

Practice 2

1. Twice a is the same as b tripled

2. Three more than b is 6 less than c

3. One half of x equals 20

4. Ten less than c is 18

5. Seven more than the product of a and 7 is one

6. The sum of x and y is four more than three times z

7. Twice a is three more than b

8. D equals 4 less than 6 times y

9. A divided by 7 is the same as 7 divided by 2

10. One-third of a equals three less than b

11. X is 5 more than y divided by 3

12. A is twice c

13. One ninth of d is the same as three more than c

14. Y is five less than six times z

15. Seven times the sum of x and 7 is 10

16. 8 more than one-half of a is six more than b

17. 5 times x plus 8 equals four less than y

18. Ten less than 3 divided by b is m

19. Four is one-half of t

20. The difference between a and 8 is the sum of b and 10

Translate Phrases and Sentences into Inequalities

KISS IT

Step 1: Identify key words

Step 2: Plug in values

Greater than >	**Less than <**
Greater Larger Exceeds More than Above Higher	Is less than Smaller Fewer than Below Better Deal
Greater than or equal to ≥ Minimum At least No less than top	**Less than or equal to ≤** Maximum At most No more than bottom

Example 5

Six plus one half of d is at least 10

Step 1: Identify the key words

Six **plus** one half **of** d is **at least** 10

addition multiplication greater than or equal to

Step 2: Plug in values

Six **plus** one half **of** d is **at least** 10

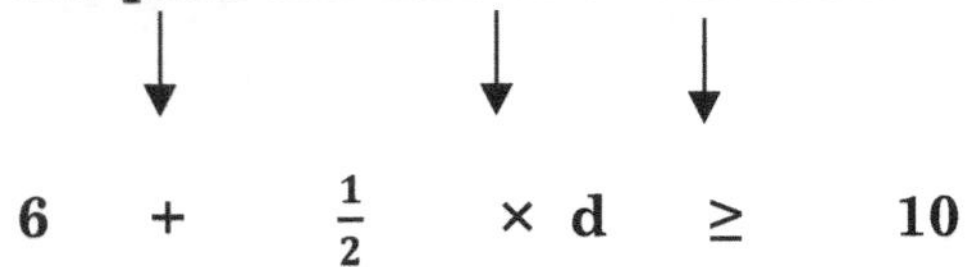

$$6 + \frac{1}{2} \times d \geq 10$$

Example 6

Five less than c is no more than six

Step 1: Identify key words

Five **less than** c is **no more** than six

$c - 5 < 6$

Practice 3

1. Three more than c is greater than 5

2. Five less than a is at most 12

3. Twice b is no more than five more than c

4. One half of x is greater than 5 less than y

5. The sum of a and b has a maximum of c

6. The difference between m and 5 exceeds 3

7. Three times a is no more than one half of b

8. The sum of c and 8 exceeds 10

9. Ten less than b is at least 25

10. The sum of x and 7 is less than 14

11. The difference between a and 8 is at most the sum of b and 10

12. Seven more than the product of a and 7 is less than one half

13. Seven less than a times b is no more than 5

14. Twice a is greater than or equal to twice b

15. One-third of x is at most 45

16. Six less than b exceeds 20

17. Five more than b is at least 2

18. Three times g minus 4 is no more than 10

19. X divided by 2 is at most 5

20. The sum of y and z is at least 23

Word Problems

KISS IT

Step 1: Identify key words

Step 2: Plug in values

Example 7

The number of girls (g) is six less than 5 times the number of boys (b) in Ms. Senat's class. Describe the number of girls in terms of (b).

Step 1: Identify key words

The number of girls (g) **is** six **less than** 5 **times** the number of boys (b) in Ms. Senat's class. Describe the number of girls (g) in terms of boys (b).

is: equal sign

less than: subtraction

times: multiplication

> In terms of is also a phrase used as an equal sign. This means that "g" is equal to an expression containing the variable "b"

Step 2: Plug in values

(g) **is** six **less than** 5 **times** the number of boys (b)

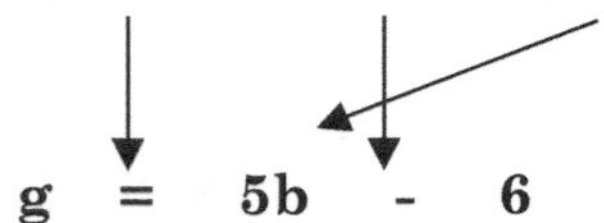

g = 5b - 6

Practice 4

1. The number of basketballs (b) that Shank can paint in one hour is ten less than one-half the number of softballs (s) she can paint in the same time. Describe the number of basketballs (b) in terms of (s).

2. Monday (m) Stephanie sold six more than three times the number of lashes she sold on Tuesday (t). Describe the number of lashes sold Monday (m) in terms of (t).

3. Brittney (b) can execute twelve more than 4 times the number high kicks (h) required for dance tryouts. Describe the number of kicks Brittney (b) can execute in terms of (h).

4. Shaunte (s) can complete thirteen more than three times the number of charts (c) required for her weekly quota. Describe Shaunte's (s) completed charts in terms of (c).

5. Vikki (v) has completed ten less than one-half the number of hours (h) required for her internship. Describe Vikki's (v) number of completed hours in terms of (h).

Example 8

Eleven more than 10 times the number of squirrels in the park is forty-two. Write an equation that can be used to solve for the number of squirrels in the park.

Step 1: Identify key words

Eleven **more than** 10 **times** the number of squirrels in the park **is** forty-two.

more than: addition

times: multiplication

is: equal sign

Step 2: Plug in values

Eleven **more than** 10 **times** the number of squirrels in the park **is** forty-two.

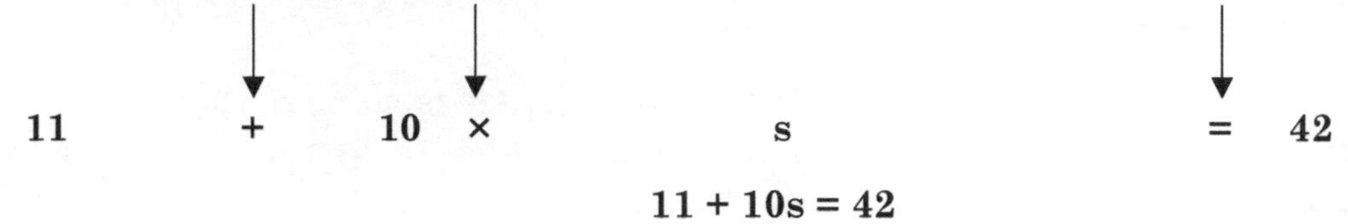

$$11 + 10s = 42$$

Practice 4

6. Adrianne is required to create ten more than three times the total number of candles ordered over the weekend due to additional orders placed after midnight. If the total number of candles ordered is 106, create an equation that illustrates the number of candle orders that need to be fulfilled.

7. Sabrina is required to register at least 100 students each day during her shift. She registered two more than two times the number of students required. Create an inequality illustrating this scenario.

8. Shameika completed 50 more than 2 times the number of hours required for her nursing internship. She is required to complete 150 hours. Create an equation to illustrate this scenario.

9. Allen's football roster has 2 less than one-half the required number of players needed for approval. If he is required to register at least 60 players, create an inequality to illustrate this scenario.

10. Monday Terrel caught twelve less than one-fourth the number of fishes he caught Sunday. Create an expression to illustrate this scenario.

Review

1. Six less than 4 times t

2. The difference between m and 5 exceeds 3

3. Three more than b is 6 less than c

4. 5 more than double c

5. The number of football players (f) is 20 more than two times the number of basketball players (b) at the local high school. Describe the number of football players (f) in terms of the number of basketball players (b).

6. The difference between s and 8 is no more than 32

7. One fourth of the sum of a and 8

8. The sum of x and y is four more than three times z

9. Five less than one-third the number of cars in the parking garage is no more than 56. If Lisa wanted to calculate the exact number of cars in the garage what inequality would she use to solve the problem?

10. Twice b is more than four less than c

11. Seven less than a times b is no more than 5

12. Nine less than half of x

13. Twice the number of cheerleaders is no more than one-third the number of football players. Write an inequality that illustrates this scenario.

14. The sum of x and y is three times z

15. Four more than 5 times the number of cats in the alley is 2. Write an equation to illustrate this scenario.

16. Four times d plus 19

17. Ten less than 3 divided by b is m

18. Esco ate five more than one-third of the number of dog treats that Toot ate. Describe the number of treats Esco (e) ate in terms of the number of treats Toot (t) ate.

19. 8 more than one-half of a is six

20. Three times g minus 4 is at least 1

Practice 1

1. $\frac{3}{x} - 10$

2. $5c + 4$

3. $\frac{1}{2}(5 + b)$

4. $\frac{1}{4}x + \frac{1}{8}y$

5. $(b - c) + 2$

6. $ab - 7$

7. $2z + 5$

8. $4t - 6$

9. $3x - 12$

10. $\frac{1}{2}x - 9$

11. $2z$

12. $\frac{1}{4}(a + 8)$

13. $4d + 19$

14. $5g + 7$

15. $y - 10$

16. $6x + 90$

17. $6d - 36$

18. $2c + 5$

19. $3b$

20. $3x - 5$

Practice 2

1. $2a = 3b$

2. $b + 3 = c - 6$

3. $\frac{1}{2}x = 20$

4. $c - 10 = 18$

5. $7a + 7 = 1$

6. $x + y = 3z + 4$

7. $2a = b + 3$

8. $d = 6y - 4$

9. $\frac{a}{7} = \frac{7}{2}$

10. $\frac{1}{3}a = b - 3$

11. $x = \frac{y}{3} + 5$

12. $a = 2c$

13. $\frac{1}{9}d = c + 3$

14. $y = 6z - 5$

15. $7(x + 7) = 10$

16. $\frac{1}{2}a + 8 = b + 6$

17. $5x + 8 = y - 4$

18. $\frac{3}{b} - 10 = m$

19. $4 = \frac{1}{2}t$

20. $a - 8 = b + 10$

Practice 3

1. $c + 3 > 5$

2. $a - 5 \leq 12$

3. $2b \leq c + 5$

4. $\frac{1}{2}x > y - 5$

5. $a + b \leq c$

6. $m - 5 > 3$

7. $3a \leq \frac{1}{2}b$

8. $c + 8 > 10$

9. $b - 10 \geq 25$

10. $x + 7 < 14$

Practice 3

11. $a - 8 \leq b + 10$

12. $7a + 7 < \frac{1}{2}$

13. $ab - 7 \leq 5$

14. $2a \geq 2b$

15. $\frac{1}{3}x \leq 45$

16. $b - 6 > 20$

17. $b + 5 \geq 2$

18. $3g - 4 \leq 10$

19. $\frac{x}{2} \leq 5$

20. $y + z \geq 23$

Practice 4

1. $b = \frac{1}{2}s - 10$

2. $m = 3t + 6$

3. $b = 4h + 12$

4. $s = 3c + 13$

5. $v = \frac{1}{2}h - 10$

6. $3x + 10 = 106$

7. $2x + 2 \geq 100$

8. $2x + 50 = 150$

9. $\frac{1}{2}x - 2 \geq 60$

10. $\frac{1}{4}x - 12$

Review

1. $4t - 6$

2. $m - 5 > 3$

3. $b + 3 = c - 6$

4. $2c + 5$

5. $f = 2b + 20$

6. $s - 8 \leq 32$

7. $\frac{1}{4}(a + 8)$

8. $x + y = 3z + 4$

9. $\frac{1}{3}c - 5 \leq 56$

10. $2b > c - 4$

11. $ab - 7 \leq 5$

12. $\frac{1}{2}x - 9$

13. $2c \leq \frac{1}{3}f$

14. $x + y = 3z$

15. $5c + 4 = 2$

16. $4d + 19$

17. $\frac{3}{b} - 10 = m$

18. $e = \frac{1}{3}t + 5$

19. $\frac{1}{2}a + 8 = 6$

20. $3g - 4 \geq 1$

BLANK PAGE

Solve Equations In One Variable

You will learn:

How to solve equations with variables on both sides

How to solve equations using the distributive property

How to solve inequalities

How to solve absolute value equations

How to solve absolute value inequalities

Study Tips

Read and study EVERY example problem.

Complete EVERY practice problem.

Check to make sure all answers are correct.

Go back to correct the questions you answered incorrectly.

If you don't receive at least an 80% on the review, go back and practice the topic.

KISS IT

Step 1: Combine like terms

Step 2: Isolate the variable

Step 3: Solve

Variables on both sides

Example 1

$3x - 4 = 2x + 10$

Step 1: Combine like terms

$$3x - 4 = 2x + 10$$
$$-2x \quad\quad -2x$$
$$1x - 4 = 10$$
$$+4 \quad = +4$$

Step 2: Isolate the variable

$$\frac{1x}{1} = \frac{14}{1}$$

Step 3: Solve

$$x = 14$$

Distributive Property

Example 2

$2\ (x + 5) = 30$

***Distribute the 2 first**

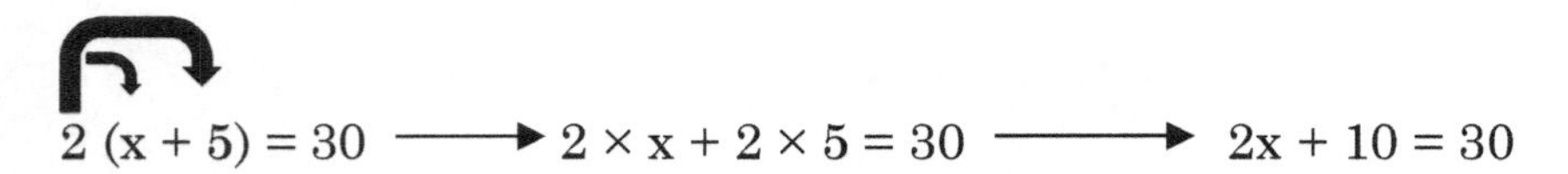

$2\ (x + 5) = 30 \longrightarrow 2 \times x + 2 \times 5 = 30 \longrightarrow 2x + 10 = 30$

Step 1: Combine like terms

$$2x + 10 = 30$$
$$-10 \quad -10$$

Step 2: Isolate the variable

$$\frac{2x}{2} = \frac{20}{2}$$

Step 3: Solve

$$x = 10$$

Practice 1

1. $2x + 4 = 8x - 8$

2. $-3x - 2 = 6x - 47$

3. $10x + 10 = 50x - 35$

4. $6x + 3 = 5x + 3$

5. $7x - 18 = 5x + 10$

6. $3x + 2 = 5x - 10$

7. $-10x + 8 = 10x - 32$

8. $1x - 8 + 3x = 12 - 1x$

9. $12 - 8x + 4x = 8 + 2x$

10. $20x + 5 + 15 = 10x + 25$

11. $3(x + 4) = 4x + 10$

12. $2(x - 3) = x + 4$

13. $3x + 5 = 4(x - 10)$

14. $4(x - 2) = 4(2x - 5)$

15. $6(x + 2) = 5(x - 2)$

16. $5(x + 2) = 18$

17. $4(x + 8) = 5(x + 2)$

18. $3(x - 9) = 3$

19. $-2(x + 2) = 3(x - 8)$

20. $4(x + 10) = 50$

Inequalities

Example 3 $3x - 3 > 8x + 17$

Step 1: Combine like terms

$$3x - 3 > 8x + 17$$
$$-8x \quad -8x$$
$$-5x - 3 > 17$$
$$+3 \quad +3$$

Step 2: Isolate the variable

$$\frac{-5x}{-5} > \frac{20}{-5}$$

Whenever you multiply or divide by a negative number, the inequality sign flips

Step 3: Solve

$$x < -4$$

Practice 2

1. $4x + 2 < 2x - 10$

2. $6x + 5 \geq 10 + 1x$

3. $3x + 2 > 15x + 22$

4. $10x + 4 < 12x - 8$

5. $5 + 3x < 6 + 8x$

6. $8(x - 2) \leq 3x + 2$

7. $-4(x - 2) > 8x + 2$

8. $3(x + 4) > 6(x - 4)$

9. $5(x - 10) \leq 10(x + 2)$

10. $2(x + 2) \geq 3(x - 1)$

Absolute Value Equations

KISS IT

Step 1: Eliminate the absolute value sign

Step 2: Rewrite the equation twice

Step 3: Solve

Example 4

$| 3x - 10 | = 11$

Step 1: Eliminate the absolute value sign

Step 2: Rewrite the equation twice

***(one positive & one negative)**

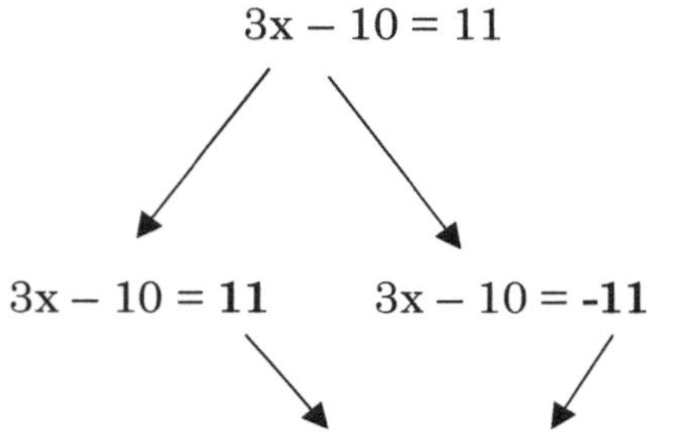

Step 3: Solve

$$
\begin{aligned}
3x - 10 &= 11 \\
+10 &= +10 \\
\hline
\frac{3x}{3} &= \frac{21}{3} \\
x &= 7
\end{aligned}
\qquad
\begin{aligned}
3x - 10 &= -11 \\
+10 &= +10 \\
\hline
\frac{3x}{3} &= \frac{-1}{3} \\
x &= -\frac{1}{3}
\end{aligned}
$$

NOTICE THE DIFFERENCE BETWEEN THE TWO VALUES: one is POSITIVE and the other is NEGATIVE

Practice 3

1. $|2x + 4| = 8$

2. $|3x + 10| = 31$

3. $|6x + 9| = 27$

4. $|5x - 2| = 18$

5. $|7x + 1| = 15$

6. $|10x + 9| = 81$

7. $|3x - 5| = 16$

8. $|12x + 10| = 62$

9. $|4x - 5| = 5$

10. $|1x + 2| = 14$

Absolute Value Inequalities

KISS IT

Step 1: Eliminate the absolute value sign

Step 2: Rewrite the equation twice

Step 3: Solve

Example 5

$|3x - 10| > 11$

Step 1: Eliminate the absolute value sign

Step 2: Rewrite the equation twice

***(one positive & one negative)**

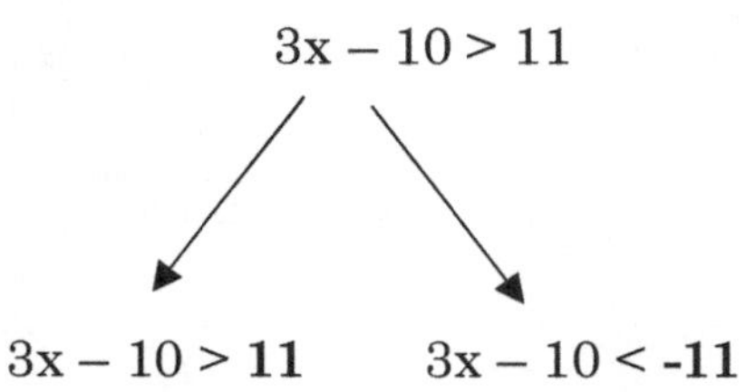

Flip the inequality sign for the second equation with the negative

Step 3: Solve

$3x - 10 > 11$

$+10 > +10$

$\frac{3x}{3} > \frac{21}{3}$

$x > 7$

$3x - 10 < -11$

$+10 < +10$

$\frac{3x}{3} < \frac{-1}{3}$

$x < -\frac{1}{3}$

Practice 4

1. $|5x + 5| < 25$

2. $|3x - 6| \leq 36$

3. $|9x - 18| > 18$

4. $|4x - 4| \geq 4$

5. $|6x + 12| < 12$

6. $|16x + 32| \leq 64$

7. $|10x + 8| \geq 12$

8. $|5x - 8| \leq 10$

9. $|25x - 11| < 11$

10. $|6x + 4| > 16$

Review

1. $4x + 10 = 5x - 8$

2. $|3x + 5| = 11$

3. $2x + 9 > 4x - 8$

4. $\frac{1}{8}(8x + 8) = 4$

5. $2(x + 8) = 4x + 10$

6. $2x + 10 \leq 15$

7. $2x + 6 + 8x = 11x + 12$

8. $|4x + 1| > 5$

9. $4(x + 2) \leq 12$

10. $\frac{1}{2}(2x + 4) = 10$

11. $|2x + 8| \leq 16$

12. $3x + 6 = 6 - 3x$

13. $\frac{1}{4}(x + 8) = \frac{1}{2}(x - 2)$

14. $3(x + 2) = 5x + 10$

15. $\frac{2}{3}x + 4 = \frac{1}{2}x - 2$

16. $10x + 14 > 24$

17. $|4x - 8| = 24$

18. $23x + 10 < -20x + 53$

19. $3x - 8 + 4x = 18$

20. $\frac{1}{3}(x - 6) = 5$

Answer Key

Practice 1

1. 2

2. 5

3. $\frac{9}{8}$

4. 0

5. 14

6. 6

7. 2

8. 4

9. $\frac{2}{3}$

10. $\frac{1}{2}$

11. 2

12. 10

13. 45

14. 3

15. -22

16. $\frac{8}{5}$

17. 22

18. 10

19. 4

20. $\frac{5}{2}$

Practice 2

1. $x < -6$

2. $x \geq 1$

3. $x < -\frac{5}{3}$

4. $x > 6$

5. $x > -\frac{1}{5}$

6. $x \leq \frac{18}{5}$

7. $x < \frac{1}{2}$

8. $x < 12$

9. $x \geq -14$

10. $x \leq 7$

Practice 3

1. 2, -6

2. 7, $-\frac{41}{3}$

3. 3, -6

4. 4, $-\frac{16}{5}$

5. 2, $-\frac{16}{7}$

6. $\frac{36}{5}$, -9

7. 7, $-\frac{11}{3}$

8. $\frac{13}{3}$, - 6

9. $\frac{5}{2}$, 0

10. 12, -16

Practice 4

1. $x < 4$, $x > -6$

2. $x \leq 14$, $x \geq -10$

3. $x > 4$, $x < 0$

4. $x \geq 2$, $x \leq 0$

5. $x < 0$, $x > -4$

6. $x \leq 2$, $x \geq -6$

7. $x \geq \frac{2}{5}$, $x \leq -2$

8. $x \leq \frac{18}{5}$, $x \geq -\frac{2}{5}$

9. $x < \frac{22}{25}$, $x > 0$

10. $x > 2$, $x < -\frac{10}{3}$

Review

1. 18

2. 2, $-\frac{16}{3}$

3. $x < \frac{17}{2}$

4. 3

5. 3

6. $x \leq \frac{5}{2}$

7. -6

8. $x > 1$, $x < -\frac{3}{2}$

9. $x \leq 1$

10. 8

11. $x \leq 4$, $x \geq -12$

12. 0

13. 12

14. -2

15. -36

16. $x > 1$

17. 8, -4

18. $x < 1$

19. $\frac{26}{7}$

20. 21

BLANK PAGE

Cluster 5 Review

You should know:

☐ How to translate sentences into expressions

☐ How to translate sentences into equations

☐ How to translate sentences into inequalities

☐ How to create equations from word problems

☐ How solve equations with variables on both sides

☐ How to solve equations using the distributive property

☐ How to solve inequalities

☐ How to solve absolute value equations

☐ How to solve absolute value inequalities

CLUSTER 5 REVIEW

1. Five more than three times x is no more than 13

a. $3x + 5 < 13$ b. $3x + 5 \leq 13$ c. $3x + 5 > 13$ d. $3x + 5 \geq 13$

2. The number of seniors is seven less than two times the number of juniors. Describe the number of seniors (s) in terms of the number of juniors (j)

a. $s = 2j - 7$ b. $s = 7 - 2j$ c. $j = 2s - 7$ d. $j = 7 - 2s$

3. $3x + 2 + 5x = 10 + 4x$

a. 2 b. 1 c. -2 d. -1

4. One half of the sum of x and 12

a. $\frac{1}{2}x + 12$ b. $\frac{1}{2} + x + 12$ c. $\frac{1}{2}(x + 12)$ d. $\frac{1}{2}(x - 12)$

5. $|2x + 4| = 16$

a. x = -6, 10 b. 6, 10 c. 6, -10 d. -6, -10

6. $4(x + 4) = 24$

a. 5 b. 2 c. 7 d. 10

7. Seventeen more than the product of a and 7 is forty-two

a. $17a + 7 = 42$ b. $7a + 17$ c. $17a + 7$ d. $7a + 17 = 42$

8. The number of girls who work out with Brandon is ten more than 3 times the number of fellas. Describe the number of girls (g) who work out with Brandon in terms of the number of fellas (f).

a. $g = 3f + 10$ b. $f = 3g + 10$ c. $g = 10f + 3$ d. $f = 10g + 3$

9. Ten more than three times the number of apples in the pantry is 65. If Gloria wanted to know the exact number of apples available, which equation would she use to solve the problem?

a. $3a + 10$ b. $10 + 3a$ c. $10 + 3a = 65$ d. $3a + 10 = 65$

10. $3x - 7 = 10x + 7$

a. 0 b. -2 c. 2 d. $\frac{1}{2}$

11. Eight less than one-half of x

a. $8 - \frac{1}{2}x$ b. $\frac{1}{2}x - 8$ c. $8 - \frac{1}{2} - x$ d. $\frac{1}{2} - x - 8$

12. five less than b is at most 23

a. $5 - b < 23$ b. $b - 5 \leq 23$ c. $b - 5 \geq 23$ d. $5 - b > 23$

13. $3x - 10 < 7$

a. $x < \frac{3}{17}$ b. $x < -1$ c. $x < 1$ d. $x < \frac{17}{3}$

14. $\frac{1}{2}(x - 4) = 12$

a. 32 b. 16 c. 28 d. 20

15. The number of nursing majors is forty-two less than four times the number of communications majors. Express the number of nursing majors (n) in terms of the number of communication majors (c).

a. $c = 4n - 42$ b. $n = 42c - 4$ c. $n = 4c - 42$ d. $4c - 42$

16. 5 times x plus 35 equals fourteen less than y

a. $5x + 35 = y - 14$ b. $5x + 35 = 14 - y$ c. $5x - 35 = y - 14$ d. $5x + 35 = y + 14$

17. $-3(x + 9) \geq 9$

a. $x \geq 12$ b. $x < -12$ c. $x \leq -12$ d. $x \leq 6$

18. Fifteen less than four times is the number of bike riders on campus is equal to 125. How many bike riders are there on campus?

a. 28 b. 33 c. 27 d. 35

19. Jakara owns ten less than three times the number of sneakers that Allen own. Express the number of sneakers Jakara owns (j) in terms of the number of sneakers Allen owns (a).

a. $j = 3a - 10$ b. $a = 3j - 10$ c. $3a - 10$ d. $3j - 10$

20. One half of the sum of the number of apples and 6 equals eight. How many apples are there?

a. 4 b. 10 c. 6 d. 8

Answer Key

1. B

2. A

3. A

4. C

5. C

6. B

7. D

8. A

9. D

10. B

11. B

12. B

13. D

14. C

15. C

16. A

17. C

18. D

19. A

20. B

Cluster 5 Review Break Down

Translate Phrases and Sentences into Expressions Equations and Inequalities

Question 1 ☐

Question 2 ☐

Question 4 ☐

Question 7 ☐

Question 8 ☐

Question 9 ☐

Question 11 ☐

Question 12 ☐

Question 15 ☐

Question 16 ☐

Question 19 ☐

Solve Equations in One Variable

Question 3 ☐

Question 5 ☐

Question 6 ☐

Question 10 ☐

Question 13 ☐

Question 14 ☐

Question 17 ☐

Question 18 ☐

Question 20 ☐

Practice Test 1

36 questions – 54 minutes

1. Alexis can type $2\frac{3}{4}$ pages in 2 hours. At this rate, how many pages can she type in 7 hours?

a. $9\frac{5}{8}$ hours b. $8\frac{19}{100}$ hours c. $8\frac{1}{4}$ hours d. 9 hours

2. Arrange the values in ascending order: $\frac{14}{5}$, 250%, $\sqrt{6}$, $\frac{7}{4}$

a. $\frac{14}{5}$, 250%, $\sqrt{6}$, $\frac{7}{4}$ b. $\frac{7}{4}$, $\sqrt{6}$, 250%, $\frac{14}{5}$ c. $\frac{7}{4}$, $\sqrt{6}$, $\frac{14}{5}$, 250% d. $\frac{14}{5}$, 250%, $\frac{7}{4}$, $\sqrt{6}$

3. Which type of correlation is depicted in the graph below?

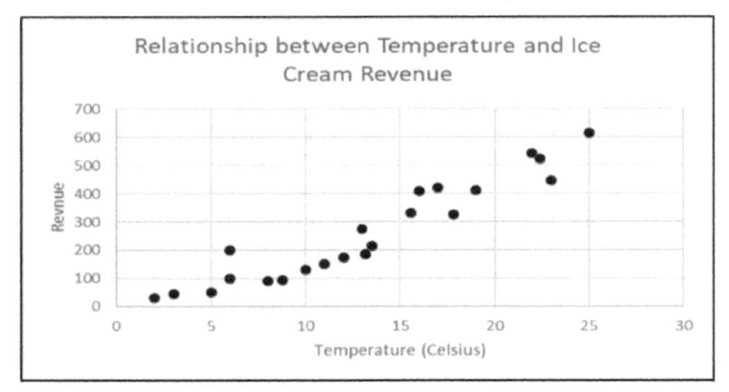

a. positive b. negative c. no correlation

4. Identify the best unit of measure for the weight of a newborn puppy

a. liters b. ounces c. tons d. milligrams

5. Based on the graph below, how many people chose Yorkshire Terriers as their favorite dog breed if 150 people were surveyed?

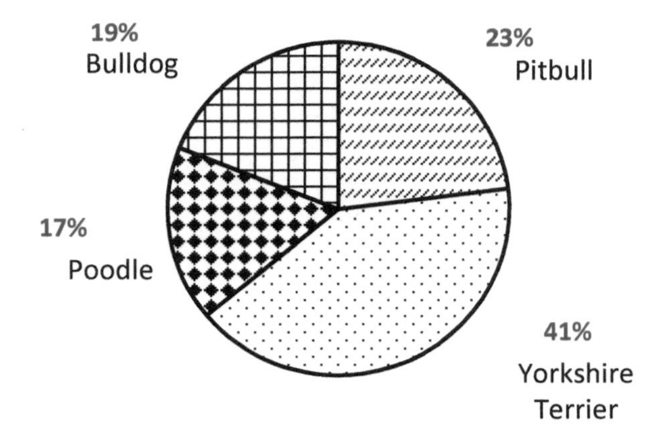

a. 41 b. 59 c. 62 d. 38

6. $2\frac{1}{4} + 3\frac{2}{3}$

a. $5\frac{37}{100}$ b. $5\frac{3}{7}$ c. $5\frac{11}{12}$ d. $6\frac{11}{12}$

7. What 17% of 182

a. 30.94 b. 17 c. 10.71 d. .093

8. Gloria receives a paycheck bi-monthly for $3,859.16. If $532.98 is withheld from each paycheck, what is her net annual salary?

a. $39,914.16 b. $52,705.68 c. $105,411.36 d. $79,828.32

9. The ratio of dogs to cats at the local pound is 5 to 6. If there are 72 cats accounted for, how many dogs should there be?

a. 71 b. 60 c. 87 d. 86

10. Calculate the median of the given data set: 34 42 21 41 26 29

a. 32.2 b. 21 c. 31.5 d. 26

11. 5.38 miles = ____________km **(1 mile = 1.609 km)**

a. 8.66 b. 3.34 c. 8.50 d. 0.30

12. The face of Adrianne's favorite watch has a circumference of 5.5 centimeters. Calculate the diameter of the watch's face. (round to the nearest hundredth)

a. 1.75 cm b. 2.75 cm c. 0.88 cm d. 3.50 cm

13. Based on the chart below, what was the average number of items in inventory during the month of May?

House of JYL Inventory Count May 2018	
Bodysuits	94
Pants	62
Jumpsuits	92
Dresses	89
Two Piece Sets	120

a. 120 b. 457 c. 91 d. 100

14. Alanah baked 50 cupcakes for her school's bake sale. If she sold 80% of her cupcakes for 50 cents each, how much money did she bring home?

a. $25 b. $20 c. $40 d. $50

15. Allen's favorite football team won 60% of their games this season. If they played 30 games this season, how many games did they lose?

a. 40 b. 18 c. 30 d. 12

16. $3x - 6 = 5x - 10$

a. 2 b. -2 c. 8 d. -8

17. Malik cut a piece of plywood into 5 separate pieces: $1\frac{1}{2}$ ft $3\frac{2}{3}$ ft 2 ft $1\frac{3}{4}$ ft $\frac{9}{3}$ ft. What is the total length of the plywood?

a. $10\frac{11}{12}$ b. $11\frac{11}{12}$ c. $7\frac{15}{12}$ d. 11

18. David can paint a room in 4 hours, and Jessica can paint a room in 5 hours. If they work together, how long will it take them to paint a room?

a. 2.2 hours b. 4.5 hours c. 0.45 hours d.

19. 78.65 km = _______________ mm

a. 786.5 b. 78,650 c. 78,650,000 d. 786,500

20. Convert $\frac{18}{43}$ to a percent to the nearest hundredth

a. .42% b. 41.86% c. 2.39% d. 238.89%

21. The local hair store has a bundle deal that includes 5 bundles for $450. At this same rate, how much would two bundles cost?

a. $180 b. $225 c. $200 d. $175

22. $\frac{2}{9} \div \frac{3}{4}$

a. $\frac{1}{6}$ b. $\frac{3}{4}$ c. $\frac{8}{27}$ d. $\frac{27}{8}$

23. The ratio of base to height on a triangle is 2:3. Identify an equivalent ratio.

a. 3:2 b. 8:15 c. 4:5 d. 6:9

24. 123 inches = ____________________ yards

a. $3\frac{5}{12}$ b. $10\frac{1}{4}$ c 1,476 d. $4\frac{5}{12}$

25. Last week, Shank hand painted 12 custom denim jackets. This week she painted 8 more than last week. What is the percent increase in the number of jackets she painted?

a. 20% b. 8% c. 67% d. 40%

26. The sum of x and y is four more than three times z

a. $x + y + 4 + 3z$ b. $x + y > 3z + 4$ c. $x - y = 3z - 4$ d. $x + y = 3z + 4$

27. On Monday Kevin's available balance was $98,934.65. Tuesday, he made a purchase at the mall for $582.98. Wednesday, he wrote a check for monthly bills totaling $3,843.89. Thursday, his paycheck ($9,689.76) was deposited. What was his balance on Friday?

a. $104,197.54 b. $104,179.54 c. $113,051.28 d. $113,510.28

28. A bag contains 12,435 marbles. If one jar can hold 75 marbles, how many jars are needed to hold the total number of marbles?

a. 75 b. 166 c. 12,435 d. 160

29. What type of distribution is displayed in the graph below?

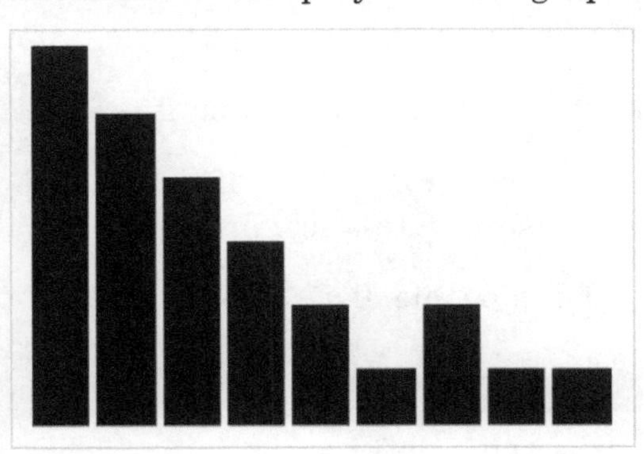

a. positively skewed b. negatively skewed c. normal d. uniform

30. Estimate the length of a loaf of bread

a. 0.3 in b. 0.3 yards c. 0.3 lbs. d. 0.3 cm

31. 2.56 g = _______________ mg

a. 25.6 b. 256 c. 2,560 d. 25,600

32. $|2x - 8| = 4$

a. 6,6 b. 2, -2 c. -6, 2 d. 6, 2

33. An empty lot measures 50 feet long and 20 feet wide. The owner wants to build a fence around the property, how much fencing does he need to purchase?

a. 140 ft b. 1,000 ft c. 70 ft d. 500 ft

34. Vikki purchased 10 pairs of shoes in one week. At this rate, how many pairs of shoes will she have purchased in 3 months?

a. 30 b. 60 c. 90 d. 120

35. Six more than one-half the number of cheerleaders is no more than 24. If Coach Edwards wants to calculate the exact number of cheerleaders on the squad, what inequality would she use to solve the problem?

a. $\frac{1}{2}x + 6 \leq 24$ b. $\frac{1}{2}x + 6 < 24$ c. $\frac{1}{2}x + 6 > 24$ d. $\frac{1}{2}x + 6 \geq 24$

36. As people age, their immune system weakens. Identify the dependent variable in the given scenario.

a. People b. Age c. Immune System d. Their

Answers and Explanations

1. A. $\frac{\frac{11}{4}\text{ pages}}{2\text{ hours}} = \frac{?\text{ pages}}{7\text{ hours}}$. After cross multiplying $(\frac{11}{4} \times 7) \div 2 = 9\frac{5}{8}$ hours

2. B. After converting the values, you have: 2.8 2.50 2.45 1.75. Ascending means from least to greatest. Therefore, the correct order is $\frac{7}{4}$, $\sqrt{6}$, 250%, $\frac{14}{5}$

3. A. The data is slanted upward from left to right, thus making it positive.

4. B. Liters measure volume. Tons is too large (1 ton = 2,000 lbs.). Milligrams are too small. Ounces is the most appropriate unit of measure in this scenario.

5. C. 41% chose Yorkshire terriers out of 150 people. 0.41 × 150 = 61.5 ⟶ 62

6. C. $\frac{9}{4} + \frac{11}{3} = \frac{27}{12} + \frac{44}{12} = \frac{71}{12} = 5\frac{11}{12}$

7. A. The word "of" means multiply. 0.17 × 182 = 30.94

8. D. \$3,859.16 - \$532.98 = \$3,326.18 each paycheck. \$3,326.18 × 2 = \$6,652.36 each month. \$6,652.36 × 12 = \$79,828.32 each year.

9. B. $\frac{5\text{ dogs}}{6\text{ cats}} = \frac{?\text{ dogs}}{72\text{ cats}}$. After cross multiplying (72×5) ÷ 6 = 60 dogs.

10. C. Place the numbers in order from least to greatest: 21 26 29 34 41 42. There are two number in the middle: 29 and 34. The average (29+34) ÷2 = 31.5.

11. A. When converting from one unit to another you need to set up a ratio. 5.38 miles × $\frac{1.609\text{ km}}{1\text{ mile}}$. Because the two numbers are at the same level, you should multiply which equals 8.66 km.

12. A. C = πd. 5.5 = 3.14d. 1.75 = d

13. C. Total number of items (94+62+92+89+120) = 457. 457 ÷ 5 = 91.4

14. B. 80% of 50 = 0.80 × 50 = 40 cupcakes at 0.50 each (40 × 0.50) = \$20.

15. D. 60% of 30 games = 18 games won. 30 – 18 = 12 games lost.

16. A. 3x – 6 = 5x – 10. Add 10 to both sides, 3x + 4 = 5x. Subtract 3x from both sides, 4 = 2x. Divide both sides by 2, 2 = x.

17. B. Total length means addition. Convert each mixed number into an improper fraction (save the whole number (2) for later): $\frac{3}{2} + \frac{11}{3} + \frac{7}{4} + \frac{9}{3} = 9\frac{11}{12} + 2 = 11\frac{11}{12}$

18. A. (4 × 5) / (4 + 5) = 20/9 = 2.2 hours

19. C. When converting from kilometers to millimeters you need to move the decimal 6 spaces to the right.

20. B. Change the fraction to a decimal first 18 ÷ 43 = .4186. Then convert to a percentage, .4186 × 100 = 41.86%

21. A. $\frac{5\text{ bundles}}{\$450} = \frac{2\text{ bundles}}{?\,\$}$. After cross multiplying (450×2) ÷ 5 = $180

22. C. $\frac{2}{9} \div \frac{3}{4} \longrightarrow \frac{2}{9} \times \frac{4}{3} = \frac{8}{27}$

23. D. The ratio 2:3 is equivalent to 6:9. When you cross multiply the two ratios $\frac{2}{3} = \frac{6}{9}$, both sides are equal…18=18.

24. A. When converting from one unit to another you need to set up a ratio.

123 inches × $\frac{1\text{ yard}}{36\text{ in}}$. Because the two numbers are at different levels, you should divide thus giving you $3\frac{5}{12}$.

25. C. 12 + 8 = 20 jackets this week. Percentage Increase = $\frac{20-12}{12}$ = 67%

26. D. The sum of x and y means addition, is means equals, four more means addition and 3 times x means multiplication.

27. A. ($98,934.65 - $582.98 - $3,843.89) + $9,689.76 = $104,197.54

28. B. $\frac{1\text{ jar}}{75\text{ marbles}} = \frac{?\text{ jars}}{12{,}435\text{ marbles}}$. After cross multiplying (1×12,435) ÷ 75 = 165.8 $\longrightarrow$ 166

29. A. When graphs fall from left to right, it represents a positively skewed.

30. B. 0.3 in. is too short, 0.3 lbs. measures weight, and 0.3 cm is too short. The most appropriate estimate is 0.3 yards.

31. C. When converting from grams to milligrams you need to move the decimal 3 spaces to the right.

32. D. Rewrite the absolute value as two new equations: 2x – 8 = 4 and 2x – 8 = -4. 2x – 8 = 4, add 8 to both sides and then divide by 2, x = 6. 2x – 8 = -4, add 8 to both sides and then divide by 2, x = 2.

33. A. Fencing around the lot represents the perimeter. P = 2l + 2w, P = 2(50) + 2(20) = 140 ft

34. D. $\frac{10\text{ pairs}}{1\text{ week}} = \frac{?\text{ pairs}}{12\text{ weeks}}$. After cross multiplying (12×10) ÷ 1 = 120

35. A. Six more means addition, one-half "of" means multiplication, and no more than means less than or equal to.

36. C. Immune systems weaken as people are aging, which means its depending on the age. This makes it the dependent variable.

Practice Test 1 Breakdown

Apply estimation strategies...

Question 4 ☐

Question 30 ☐

Convert among non-negative fractions...

Question 20 ☐

Compare and order rational numbers

Question 2 ☐

Convert within and between standard...

Question 11 ☐

Question 19 ☐

Question 24 ☐

Question 31 ☐

Solve problems involving ratios...

Question 9 ☐

Question 23 ☐

Solve problems involving proportions

Question 1 ☐

Question 34 ☐

Question 21 ☐

Solve problems involving percentages

Question 7 ☐

Question 15 ☐

Question 25 ☐

Interpret relevant information...

Question 5 ☐

Question 13 ☐

Evaluate the information in tables...

Question 10 ☐

Question 29 ☐

Calculate geometric quantities

Question 12 ☐

Question 33 ☐

Perform arithmetic operations ration...

Question 6 ☐

Question 22 ☐

Solve problems rational numbers

Question 8 ☐

Question 14 ☐

Question 17 ☐

Question 18 ☐

Question 27 ☐

Question 28 ☐

Translate phrases and sentences into...

Question 26 ☐

Question 35 ☐

Solve equations in one variable

Question 16 ☐

Question 32 ☐

Explain the relationship between variables

Question 3 ☐

Question 36 ☐

Practice Test 2

36 questions – 54 minutes

1. $\sqrt{13}$ = (to the nearest hundredth)

a. 3.60 b. 3.61 c. 3.606 d. 3.52

2. There are 22 girls and 42 boys in Coach Butler's physical education class. What is the ratio of girls to the total number of students in class?

a. 11:32 b. 11:21 c. 32:11 d. 22:42

3. Which type of correlation is depicted in the graph below?

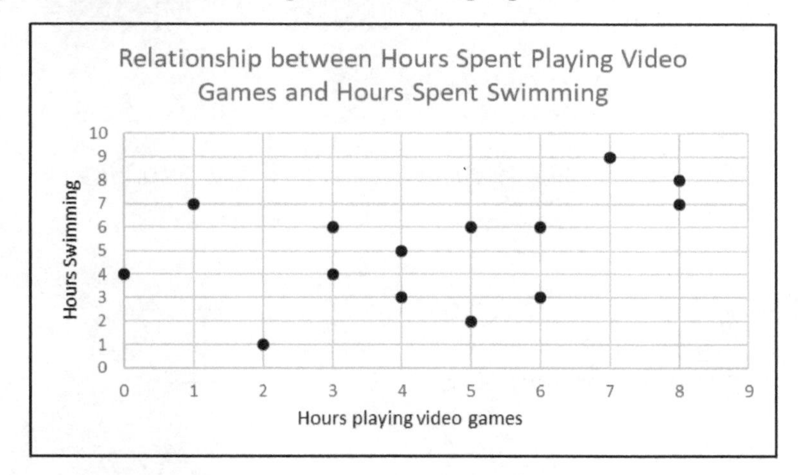

a. positive b. negative c. no correlation

4. 45°F = ______________ °C **(C =** $\frac{5}{9}$ **(F – 32))**

a. 7.2° b. 17.8° c. 7.8° d. 45°

5. Calculate the mean of the given data set: 23 34 10 45 34

a. 34 b. 35 c. 146 d. 29

6. There are 23 students for every teacher at the local high school. If the school has a student population of 1,127, how many teachers are there?

a. 49 b. 23 c. 25,921 d. 45

7. Identify the fraction that is equivalent to 68%

a. $\frac{3}{5}$ b. $\frac{17}{25}$ c. $\frac{25}{17}$ d. $\frac{86}{100}$

8. Based on the graph below, which student has the highest average test score?

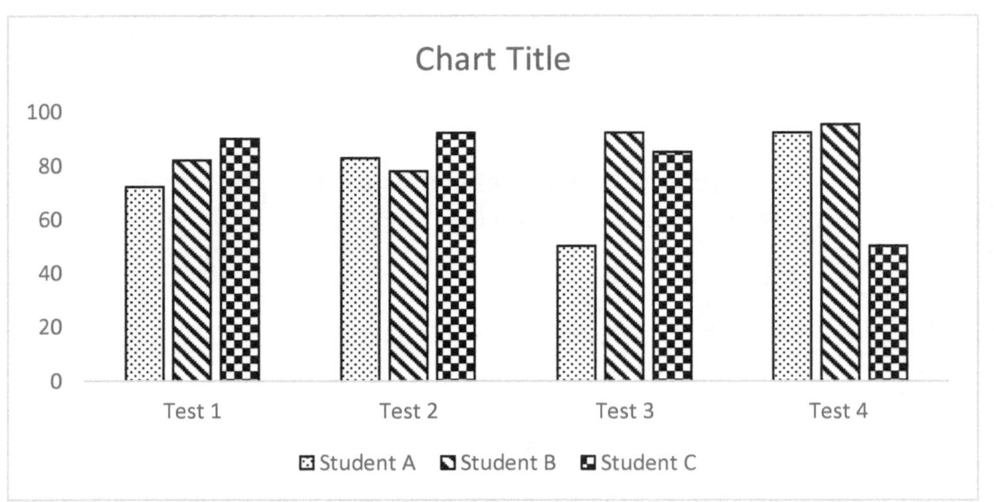

a. Student A b. Student B c. Student C d. Student D

9. $8\frac{2}{9} \div 2\frac{2}{3}$

a. $4\frac{1}{3}$ b. $6\frac{4}{12}$ c. $21\frac{25}{27}$ d. $3\frac{1}{12}$

10. Identify the largest value from the data set: -4 -10 -3 -0.5 $-\frac{2}{3}$

a. -10 b. -0.5 c. $-\frac{2}{3}$ d. -4

11. $2(x - 8) = 24$

a. 2 b. 16 c. -2 d. 20

12. 15 is what percent of 39?

a. 5.85% b. 2.60% c. 38% d. 26%

13. For every female executive at Landing & Co. there are 3 male executives. If there are 24 male executives in all, how many female executives are there?

a. 8 b. 22 c. 24 d. 10

14. 3.45 g = ___________ kg

a. 34.5 b. 0.345 c. 0.00345 d. 345

15. Based on the graph below, what is the relationship between the mean and median?

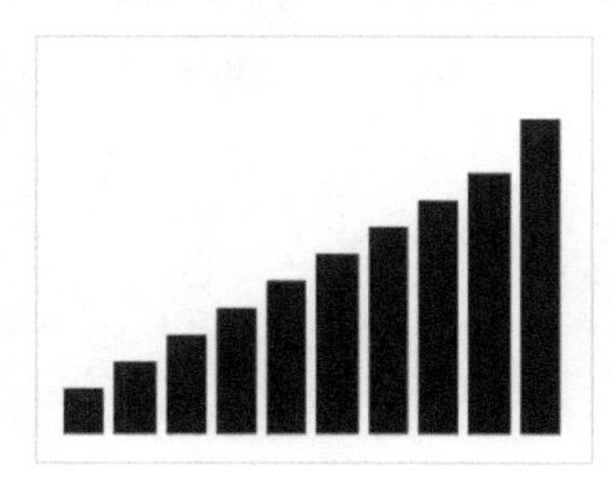

a. mean < median   b. mean > median   c. mean = median   d. mean = range

16. Jennifer purchased 3 blouses ($12.99 each), 2 skirts ($19.99 each), and 4 belts ($9.99) each. If the sales tax is 5%, how much will she pay in taxes?

a. $118.91   b. $5.95   c. $124.86   d. $5.00

17. A pizza has a diameter of 14 inches. What is its area in terms of pi?

a. 14π   b. 7π   c. 196π   d. 49π

18. Estimate the mass of a small light bulb

a. 34 mg   b. 34 g   c. 34 hg   d. 34 kg

19. Jeffery is paid $2,459 monthly. If $149.32 is withheld from his check bi-monthly, what is his net annual salary?

a. $25,924.32   b. $27,716.16   c. $31,299.84   d. $33,091.68

20. Justine can bake 148 cookies in $3\frac{1}{4}$ hours. She just received an order for 30 dozen cookies, how long will it take her to bake them?

a. 0.66 hours   b. 7.6 hours   c. 7.9 hours   d. 0.76 hours

21. Based on the graph below, what percentage of the students surveyed prefer Hip-Hop and R&B music?

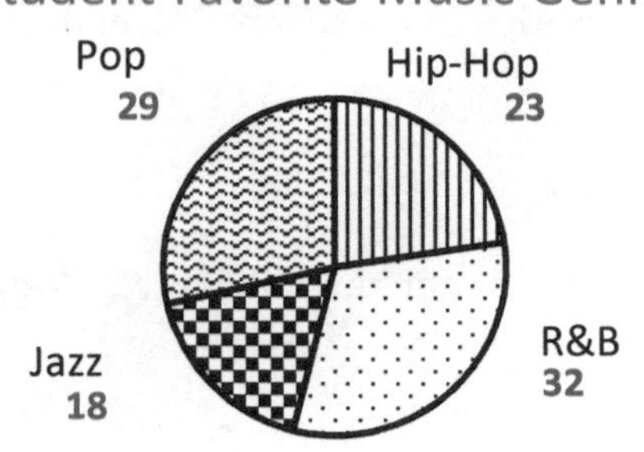

a. 23%   b. 32%   c. 51%   d. 54%

22. Raekia can complete a standard family photo shoot in $2\frac{1}{2}$ hours. Lance can complete the same photo shoot in $1\frac{3}{4}$ hours. If decide to collaborate, how long will it take them to complete the photo shoot?

a. 1.03 hours b. 0.97 hours c. 2.1 hours d. 1.30 hours

23. If the area of a triangle is 36 in^2 and it has a height of 6 inches, what is the length of its base?

a. 5.14 in b. 6 in c. 12 in d. 30 in

24. $10\frac{2}{5} - 3\frac{1}{3}$

a. $7\frac{1}{2}$ b. $7\frac{1}{15}$ c. $\frac{42}{2}$ d. $7\frac{1}{5}$

25. What is the total mass of a cake, with one layer of strawberry filling and two layers of vanilla filling?

Cake Fillings By Layer	
Vanilla	$\frac{1}{2}$ lb.
Chocolate	$\frac{1}{2}$ lb.
Strawberry	$\frac{3}{4}$ lb.
Banana Cream	$\frac{1}{4}$ lb.

a. $\frac{3}{4}$ lb. b.1 lb. c. $1\frac{3}{4}$ lb. d. $\frac{1}{2}$ lb.

26. Arrange the values in descending order: $\sqrt{36}$ $5\frac{7}{8}$ 650% 5.987

a. $\sqrt{36}$, $5\frac{7}{8}$, 650%, 5.987 b. 650%, $\sqrt{36}$, 5.987, $5\frac{7}{8}$ c. 650%, $\sqrt{36}$, $5\frac{7}{8}$, 5.987

27. The number of sophomores (s) is twelve less than one-third of the number of juniors (j) who qualified for the national academic competition. Describe the number of sophomores (s) in terms of the number of juniors (j).

a. $s = \frac{1}{3}j - 12$ b. $j = \frac{1}{3}s - 12$ c. $s = 12j - \frac{1}{3}$ d. $j = 12s - \frac{1}{3}$

28. The ratio of male teachers to female teachers in Peach County is 1:10. If there are 580 female teachers, how many male teachers are there?

a. 571 b. 5,800 c. 85 d. 58

29. As the number of hurricanes increases, bottled water sales increase. Identify the independent variable.

a. Water sales b. Hurricanes c. Increases d. Number

30. Based on the chart below, how much did Katie spend during her trip to the grocery store if she purchased 2 lbs. of grapes, 1 loaf of bread, and 2 gallons of milk?

Paul's Grocery Depot	
Fruit	$2.89/lb.
Bread	$1.99/loaf
Water	$3.99/case
Dairy	$4.99/gallon
Meat	$9.99/lb.

a. $9.87 b. $12.76 c. $15.75 d. $17.75

31. $5\frac{2}{5}$ yards = ________________ feet

a. 16.2 b. 15 c. 194.4 d. 180

32. $\frac{5}{9} \times 1\frac{2}{7}$

a. $1\frac{10}{63}$ b. $\frac{5}{7}$ c. $\frac{35}{81}$ d. $1\frac{7}{16}$

33. The difference between x and y is no more than 29

a. $x - y < 29$ b. $x - y > 29$ c. $x - y \leq 29$ d. $x - y \geq 29$

34. Sarah owns 125 pairs of shoes. If 40% of these shoes are high heels, how many high heels does she own?

a. 40 b. 60 c. 75 d. 50

35. $2a - 5 + 4a = 10 + 1a$

a. 3 b. $\frac{15}{7}$ c. 5 d. $\frac{7}{15}$

36. $\frac{2}{3} + \frac{3}{5} + \frac{1}{2}$

a. $1\frac{23}{30}$ b. $\frac{6}{10}$ c. $\frac{6}{30}$ d. $1\frac{6}{10}$

Answers and Explanations

1. B. $\sqrt{13} = 3.60555 \longrightarrow 3.61$

2. A. 22 girls. Total students (22 + 42) = 64. 22 girls: 64 total students $\longrightarrow$ 22:64 = 11:32

3. C. No relationship between the number of hours swimming and hours playing video games. The data shows no particular pattern from left to right.

4. A. $(C = \frac{5}{9}(F - 32))$. $(C = \frac{5}{9}(45 - 32)) \longrightarrow (C = \frac{5}{9}(13)) \longrightarrow 7.2° C$

5. D. 23 + 34 + 10 + 45 + 34 = 146 ÷ 5 = 29.2

6. A. $\frac{\text{23 students}}{\text{1 teacher}} = \frac{\text{1,127 students}}{\text{? teachers}}$. After cross multiplying (1×1,127) ÷ 23 = 49 teachers.

7. B. 68% $\longrightarrow$ 0.68 (two numbers after the decimal so put over 100) $\longrightarrow \frac{68}{100} = \frac{17}{25}$

8. B. Student A (70 + 80 + 40 + 90 = 280 ÷ 4 = 70) Student B (80 + 75 + 90 + 95 = 340 ÷ 4 = 85) Student C (90 + 90 + 80 + 40 = 300 ÷ 75). Student B has the highest average of 85.

9. D. $8\frac{2}{9} \div 2\frac{2}{3} \longrightarrow \frac{74}{9} \times \frac{3}{8} = \frac{222}{72} = \frac{37}{12} = 3\frac{1}{12}$

10. B. When comparing negative numbers, the "smallest" number holds the largest value. In this case, -0.5 is the smallest number, thus giving it the largest value.

11. D. Distribute the 2 to the "x" an "8". 2x – 16 = 24. Add 16 to both sides. 2x = 40. Divide both sides by 2. x = 20.

12. C. $\frac{15}{39} = \frac{}{100}$. After cross multiplying (15 × 100) ÷ 39 = 38%.

13. A. $\frac{\text{1 female exec}}{\text{3 male execs}} = \frac{\text{? female execs}}{\text{24 male execs}}$. After cross multiplying (1×24) ÷ 3 = 8 female executives.

14. C. When converting from grams to kilograms you need to move the decimal 3 spaces to the left.

15. A. This graph is skewed to the left which means the mean < median.

16. B. (\$12.99 × 3 blouses) + (\$19.99 × 2 skirts) + (\$9.99 × 4 belts) = \$118.91 × 0.05 = \$5.95

17. D. $A = \pi r^2$. Since the diameter is 14, the radius is 7. $A = 7^2\pi$. $A = 49\pi$

18. B. Milligrams is too small. Hectograms and Kilograms are too large. Grams are the most appropriate.

19. A. (\$2,459 – (\$149.32 × 2)) = \$2,160.36 × 12 = \$25,924.32

20. C. 30 dozen needs to be converted to individual cookies (30 × 12) = 360 cookies. $\frac{\text{148 cookies}}{\text{3.25 hours}} = \frac{\text{360 cookies}}{\text{? hours}}$. After cross multiplying (360 × 3.25) ÷ 148 = 7.9 hours.

21. D. (29 + 18 + 23 + 32) = 102 total students. (23 hip-hop + 32 r&b) = 55 students. $\frac{55}{102} = .539 \longrightarrow 54\%$

22. A. (2.5 × 1.75) / (2.5 + 1.75) = 4.375/4.25 = 1.03 hours

23. C. Area = $\frac{\text{base x height}}{2}$. 36 = $\frac{\text{base x 6}}{2}$ = 72 = base × 6. 72 ÷ 6 = 12 inches.

24. B. $\frac{52}{5} - \frac{10}{3} = \frac{156}{15} - \frac{50}{15} = \frac{106}{15} = 7\frac{1}{15}$

25. C. One layer of strawberry filling ($\frac{3}{4}$). Two layers of vanilla filling ($\frac{1}{2} + \frac{1}{2}$ = 1). 1 + $\frac{3}{4}$ = 1$\frac{3}{4}$

26. B. After converting the values, you have: 6 5.875 6.5 5.987. Descending means from greatest to least. Therefore, the correct order is 650% $\sqrt{36}$ 5.987 5$\frac{7}{8}$

27. A. "is" means equals, "less than" means subtraction, "of" means multiplications.
s = $\frac{1}{3}$j – 12

28. D. $\frac{\text{1 male}}{\text{10 females}} = \frac{\text{? males}}{\text{580 females}}$. After cross multiplying (1 × 580) ÷ 10 = 58 males.

29. B. The number of hurricanes impacts bottled water sales, therefore making it the independent variable.

30. D. (2 × \$2.89) + (1 × \$1.99) + (2 × \$4.99) = \$17.75
31. A. When converting from one unit to another you need to set up a ratio. 5$\frac{2}{5}$ yards × $\frac{\text{3 feet}}{\text{1 yard}}$. Because the two numbers are at the same level, you should multiply which equals 16.2 feet.

32. B. $\frac{5}{9} \times 1\frac{2}{7} \longrightarrow \frac{5}{9} \times \frac{9}{7} = \frac{45}{63} = \frac{5}{7}$
33. C. Difference means subtract, and no more than means less than or equal to. $x - y \leq 29$
34. D. 40% of 125 ⟶ 0.40 × 125 = 50 high heels
35. A. 2a – 5 + 4a = 10 + 1a. Combine like terms on the left side of the equation (2a + 4a = 6a) ⟶ 6a – 5 = 10 + 1a. Subtract 1a from both sides ⟶ 5a – 5 = 10. Add 5 to both sides ⟶ 5a = 15. Divide both sides by 5 ⟶ a = 3.

36. A. $\frac{2}{3} + \frac{3}{5} + \frac{1}{2}. \frac{2}{3} + \frac{3}{5} = \frac{19}{15} \longrightarrow \frac{19}{15} + \frac{1}{2} = \frac{53}{30} = 1\frac{23}{30}$

Practice Test 2 Breakdown

Apply estimation strategies...

Question 1 ☐

Question 18 ☐

Convert among non-negative fractions...

Question 7 ☐

Compare and order rational numbers

Question 10 ☐

Question 26 ☐

Convert within and between standard...

Question 4 ☐

Question 14 ☐

Question 31 ☐

Solve problems involving ratios...

Question 2 ☐

Question 28 ☐

Solve problems involving proportions

Question 6 ☐

Question 13 ☐

Question 20 ☐

Solve problems involving percentages

Question 12 ☐

Question 16 ☐

Question 34 ☐

Explain the relationship between variables

Question 3 ☐ Question 29 ☐

Interpret relevant information...

Question 8 ☐

Question 21 ☐

Evaluate the information in tables...

Question 5 ☐

Question 15 ☐

Calculate geometric quantities

Question 17 ☐

Question 23 ☐

Perform arithmetic operations ration..

Question 9 ☐

Question 24 ☐

Question 32 ☐

Question 36 ☐

Solve problems rational numbers

Question 19 ☐

Question 22 ☐

Question 25 ☐

Question 30 ☐

Translate phrases and sentences...

Question 27 ☐

Question 33 ☐

Solve equations in one variable

Question 11 ☐

Question 35 ☐